住房和城乡建设部"十四五"规划教材
高等学校土木工程专业国际化人才培养全英文系列教材

Seismic Design of Building Structures (2nd Edition)

建筑结构抗震设计(第二版)

周 颖 主 编
鲁懿虬 吴 浩 副主编

中国建筑工业出版社
CHINA ARCHITECTURE & BUILDING PRESS

图书在版编目(CIP)数据

建筑结构抗震设计 = Seismic Design of Building Structures：英文 / 周颖主编；鲁懿虬，吴浩副主编. —2版. — 北京：中国建筑工业出版社，2023.1

住房和城乡建设部"十四五"规划教材. 高等学校土木工程专业国际化人才培养全英文系列教材

ISBN 978-7-112-28345-3

Ⅰ.①建… Ⅱ.①周…②鲁…③吴… Ⅲ.①建筑结构-防震设计-高等学校-教材-英文 Ⅳ.①TU352.104

中国国家版本馆CIP数据核字(2023)第017596号

责任编辑：仕　帅　吉万旺
文字编辑：卜　煜
责任校对：董　楠

住房和城乡建设部"十四五"规划教材
高等学校土木工程专业国际化人才培养全英文系列教材
Seismic Design of Building Structures（2nd Edition）
建筑结构抗震设计（第二版）
周　颖　主　编
鲁懿虬　吴　浩　副主编
*
中国建筑工业出版社出版、发行(北京海淀三里河路9号)
各地新华书店、建筑书店经销
北京科地亚盟排版公司制版
北京市密东印刷有限公司印刷
*

开本：787毫米×1092毫米　1/16　印张：13½　字数：322千字
2023年2月第二版　　2023年2月第一次印刷
定价：**42.00**元（赠教师课件）
ISBN 978-7-112-28345-3
(40684)

版权所有　翻印必究
如有印装质量问题，可寄本社图书出版中心退换
(邮政编码　100037)

Introduction

This textbook introduces the fundamental knowledge of seismic design of building structures, from seismology to earthquake engineering. It introduces the basic concepts of seismic design with typical engineering examples in China to facilitate readers to understand the concepts in the design codes and its applications. This book can serve as a guide for instructors, graduate students and practicing engineers. We hope it will contribute to the teaching of earthquake engineering and its applications to practice, as well as to the formulation and evolution of research programs.

内容提要

本教材介绍了建筑结构抗震设计的基础知识，内容涵盖地震学和地震工程。教材主要介绍了抗震设计的基本概念，结合国内的典型工程实例，帮助读者理解抗震设计规范中的重要概念及其工程应用。本教材可为高校地震工程教师、研究生和广大工程师提供指南，希望它有助于地震工程的教学和研究以及其在实际工程中的应用。

为了更好地支持相应课程的教学，我们向采用本书作为教材的教师提供课件，有需要者可与出版社联系。邮箱：jckj@cabp.com.cn，电话：（010）58337285，建工书院http://edu.cabplink.com。

出版说明

党和国家高度重视教材建设。2016年，中办国办印发了《关于加强和改进新形势下大中小学教材建设的意见》，提出要健全国家教材制度。2019年12月，教育部牵头制定了《普通高等学校教材管理办法》和《职业院校教材管理办法》，旨在全面加强党的领导，切实提高教材建设的科学化水平，打造精品教材。住房和城乡建设部历来重视土建类学科专业教材建设，从"九五"开始组织部级规划教材立项工作，经过近30年的不断建设，规划教材提升了住房和城乡建设行业教材质量和认可度，出版了一系列精品教材，有效促进了行业部门引导专业教育，推动了行业高质量发展。

为进一步加强高等教育、职业教育住房和城乡建设领域学科专业教材建设工作，提高住房和城乡建设行业人才培养质量，2020年12月，住房和城乡建设部办公厅印发《关于申报高等教育职业教育住房和城乡建设领域学科专业"十四五"规划教材的通知》（建办人函〔2020〕656号），开展了住房和城乡建设部"十四五"规划教材选题的申报工作。经过专家评审和部人事司审核，512项选题列入住房和城乡建设领域学科专业"十四五"规划教材（简称规划教材）。2021年9月，住房和城乡建设部印发了《高等教育职业教育住房和城乡建设领域学科专业"十四五"规划教材选题的通知》（建人函〔2021〕36号）。为做好"十四五"规划教材的编写、审核、出版等工作，《通知》要求：(1) 规划教材的编著者应依据《住房和城乡建设领域学科专业"十四五"规划教材申请书》（简称《申请书》）中的立项目标、申报依据、工作安排及进度，按时编写出高质量的教材；(2) 规划教材编著者所在单位应履行《申请书》中的学校保证计划实施的主要条件，支持编著者按计划完成书稿编写工作；(3) 高等学校土建类专业课程教材与教学资源专家委员会、全国住房和城乡建设职业教育教学指导委员会、住房和城乡建设部中等职业教育专业指导委员会应做好规划教材的指导、协调和审稿等工作，保证编写质量；(4) 规划教材出版单位应积极配合，做好编辑、出版、发行等工作；(5) 规划教材封面和书脊应标注"住房和城乡建设部'十四五'规划教材"字样和统一标识；(6) 规划教材应在"十四五"期间完成出版，逾期不能完成的，不再作为《住房和城乡建设领域学科专业"十四五"规划教材》。

住房和城乡建设领域学科专业"十四五"规划教材的特点：一是重点以修订教育部、住房和城乡建设部"十二五""十三五"规划教材为主；二是严格按照专业标准规范要求编写，体现新发展理念；三是系列教材具有明显特点，满足不同层次和类型的学校专业教学要求；四是配备了数字资源，适应现代化教学的要求。规划教材的出版凝聚了作者、主审及编辑的心血，得到了有关院校、出版单位的大力支持，教材建设管理过程有严格保障。希望广大院校及各专业师生在选用、使用过程中，对规划教材的编写、出版质量进行反馈，以促进规划教材建设质量不断提高。

住房和城乡建设部"十四五"规划教材办公室
2021年11月

Preface (2nd Edition)

The textbook *Seismic Design of Building Structures* is aimed to help readers to understand the fundamental theory and methodologies of seismic design of structures, as well as to use standards to design various types of building structures that are commonly used in China.

It has been more than six years since the first edition of this textbook was published in 2016. The textbook has been widely used among undergraduate and graduate students majored in civil engineering over 30 universities in more than 10 provinces. It has been gratifying to see the first edition of this book become a main resource and reference in education of earthquake engineering in China.

After the first edition was published, several national codes or standards that are relevant to the contents of this textbook have been revised significantly and the design factors in these codes were updated accordingly. In addition, a number of extra important contents were added to the syllabus of the course "*Seismic Design of Building Structures*" in Tongji University. To make the textbook state-of-the-art, all chapters were carefully reviewed and, where indicated, contents were corrected or updated. The main updates in the second edition are as follows:

1. Two national codes/ standards were revised including *Code for Seismic Design of Buildings* (GB 50011—2010) Version 2016 and *the Chinese Seismic Intensity Scale* (GB/T 17742—2020). A new *General Code for Seismic Precaution of Buildings and Municipal Engineering* (GB 55002—2021) was issued. The relevant contents in the textbook have been thus updated in accordance with these codes/standards.

2. The design factors and formulas have been updated according to the current codes/standards, such as Table 4-13 and Table 4-14.

3. Additional teaching contents have been added to the textbook, such as the design with various floor diaphragms for masonry structures in Chapter 6.

We are grateful to Ou Weize, Wang Runze, Wang Jingxin and Cen Guohua who were devoted to the final edits. We hope the second edition will be of value to the students and engineers who wish to understand seismic design of structures and apply this for real design.

Zhou Ying, Lu Yiqiu and Wu Hao
September, 2022

第二版前言

《建筑结构抗震设计》旨在帮助读者了解结构抗震设计的基本理论和方法，根据规范设计我国常用的各类建筑结构。

该教材自 2016 年第一版出版至今已过去 6 年，教材已在 10 余个省的 30 余所大学土木工程专业的本科生和研究生中广泛使用，成为我国地震工程教学的主要参考资料之一。

教材第一版出版后的 6 年期间，结构抗震相关的国家规范或标准做了大规模修订，此外同济大学"建筑结构抗震"课程的教学大纲也增加了新的内容。为使教材保持最新，作者大幅更新了教材第二版，主要更新之处有：

1. 国家先后修订了《建筑抗震设计规范》GB 50011—2010（2016 版）、《中国地震烈度表》GB/T 17742—2020，新编了《建筑与市政工程抗震通用规范》GB 55002—2021 等国家标准，教材内容做了相应更新。

2. 根据现行国家规范和标准，将设计参数和公式做了相应更新，如表 4-13、表 4-14。

3. 将新增教学内容添加到了相应的章节中，如第 6 章考虑各类楼盖刚度的砌体结构设计。

感谢区伟泽、王润泽、王婧欣、岑国华为教材第二版校对所付出的辛勤工作。希望《建筑结构抗震设计》（第二版）对学习建筑结构抗震设计并致力于其工程应用的学生们和工程师们有所帮助。

周颖　鲁懿虬　吴浩
2022 年 9 月

Preface (1st Edition)

It is recognized that seismic design of building structures is a multi-disciplinary and highly integrated course; consequently, it is essential to work together with colleagues from several fields to make the essence of these subjects involving seismic design of structures more accessible to readers. Furthermore, the seismic design in China is based on three earthquake levels, which is unique to other countries. In view of these, the authors in Tongji University edited the book *Seismic Design of Building Structures*, based on the experiences of over 10 years' study, education and engineering practice on earthquake engineering.

This book is chief edited by Zhou Ying. The 9 chapters and contributors are:
- Chapter 1. Fundamental Knowledge of Earthquakes and Ground Motions (Shan Jiazeng)
- Chapter 2. Site, Subsoil and Foundation (Dai Kaoshan)
- Chapter 3. Structural Seismic Response of Single-Degree-of-Freedom and Multi-Degree-of-Freedom Systems (Lu Zheng)
- Chapter 4. Seismic Action and the Basic Principles of Seismic Design for Building Structures (Zhou Ying)
- Chapter 5. Seismic Design of Reinforced Concrete Building Structures (Zhou Ying)
- Chapter 6. Seismic Design of Masonry Building Structures (Dai Kaoshan)
- Chapter 7. Seismic Design of Steel Building Structures (Lu Zheng)
- Chapter 8. Seismic Design of Non-structural Elements (Jiang Jiafei)
- Chapter 9. Introduction to Seismic Isolation and Energy Dissipation for Building Structures (Zhou Ying)

The highlights of this book are the integration of theoretical description and engineering practice, focusing on the elaboration of basic concepts, and illustration of the detailed examples. We hope that it can help readers understand the basic theory and fundamental method in seismic design, as well as use the codes to do seismic design for various types of building structures.

Zhou Ying, Lu Zheng, Dai Kaoshan
Jiang Jiafei, Shan Jiazeng
August, 2016

Contents

Chapter 1　Earthquakes and Ground Motions
1.1　Causes and Types of Earthquakes　　002
　　1.1.1　Internal Structure of Earth　　002
　　1.1.2　Causes of Earthquakes　　003
　　1.1.3　Types of Earthquakes　　003
1.2　Seismic Waves and Propagation　　005
　　1.2.1　Body Waves　　005
　　1.2.2　Surface Waves　　006
　　1.2.3　Earthquake Records　　007
1.3　Earthquake Magnitude and Intensity　　007
　　1.3.1　Earthquake Magnitude　　007
　　1.3.2　Earthquake Intensity　　008
　　1.3.3　Relationship between Magnitude and Intensity　　014
1.4　Earthquake Characteristics and Hazards in China　　014
　　1.4.1　Earthquake Activity and Distribution in China　　014
　　1.4.2　Characteristics of Earthquake Activity in China　　015
　　1.4.3　Earthquake Hazards in China　　016
1.5　Seismic Fortification for Building Structures　　018
　　1.5.1　Seismic Fortification Objectives　　018
　　1.5.2　Seismic Design Methods　　019

Chapter 2　Site, Subsoil and Foundation
2.1　Site　　022
　　2.1.1　Selection of Construction Site　　022
　　2.1.2　Construction Site Classification　　023
　　2.1.3　Construction Site Categories　　024
2.2　Seismic Checking for Subsoil and Foundation　　025
　　2.2.1　Earthquake Resistance of Natural Subsoil　　025
　　2.2.2　Earthquake-Resistant Checking for Natural Subsoil and Foundation　　026
　　2.2.3　Earthquake-Resistant Checking for Pile Foundation　　026
2.3　Soil Liquefaction　　027
　　2.3.1　Cause and Damage of Soil Liquefaction　　027
　　2.3.2　Discrimination of Soil Liquefaction　　028
　　2.3.3　Mitigation of Liquefaction Hazard　　031

Chapter 3　Structural Seismic Response of Single-Degree-of-Freedom and Multi-Degree-of-Freedom Systems
3.1　Free Vibration of SDOF Systems　　035

		3.1.1	Mechanical Model and the Equation of Motion	035
		3.1.2	Undamped Free Vibration of SDOF Systems	036
		3.1.3	Damped Free Vibration of SDOF Systems	037

- 3.2 Forced Vibration of SDOF Systems under Arbitrary Loading — 039
  - 3.2.1 Instantaneous Impulse and its Free Vibration — 039
  - 3.2.2 Dynamic Response under a General Dynamic Load-Duhamel Integration — 039
- 3.3 Numerical Methods for Seismic Response of SDOF Systems — 040
  - 3.3.1 Duhamel Integration Method — 040
  - 3.3.2 Numerical Solutions for Equations of Motions — 040
- 3.4 Response Spectrum for Building Design — 042
  - 3.4.1 Basic Formula for Horizontal Earthquake Action — 043
  - 3.4.2 Earthquake Effect Coefficient—Response Spectrum in Chinese *Code for Seismic Design of Buildings* (GB 50011—2010) Version 2016 — 043
- 3.5 Nonlinear Analysis — 044
  - 3.5.1 Nonlinearity of Materials — 045
  - 3.5.2 Equations of Motion of SDOF Nonlinear Systems $K(t)$ — 045
  - 3.5.3 Solution of Nonlinear Motion Equation — 045
  - 3.5.4 Hysterestic Model — 046
- 3.6 Free Vibration of Multi-Degree-of-Freedom Systems — 047
  - 3.6.1 Equations of Motion — 047
  - 3.6.2 Vibration Properties of Multi-Degree-of-Freedom Systems — 049
  - 3.6.3 Approximation Method for the Natural Properties — 051
- 3.7 Modal Analysis Method for Multi-Degree-of-Freedom Systems — 055
- 3.8 Horizontal Seismic Effect and Response of Multi-Degree-of-Freedom Systems — 057
  - 3.8.1 Response Spectrum Method — 057
  - 3.8.2 Base Shear Method — 060
- 3.9 Time History Method — 062
  - 3.9.1 Linear Acceleration Method for Multi-Degree-of-Freedom Systems — 062
  - 3.9.2 Wilson-θ Method for Multi-Degree-of-Freedom Systems — 063
  - 3.9.3 Seismic Response Calculation for Multi-Degree-of-Freedom Nonlinear Systems — 063

Chapter 4 Earthquake Effect and Seismic Design Principles

- 4.1 Building Classification and Seismic Fortification — 066
  - 4.1.1 Seismic Fortification Classification — 066
  - 4.1.2 Seismic Fortification Criterion — 066
  - 4.1.3 Seismic Fortification Objective — 067
  - 4.1.4 Three Levels — 068
- 4.2 Seismic Conceptual Design — 070
- 4.3 Calculation Methodology of Seismic Action — 073
  - 4.3.1 General Requirements — 073
  - 4.3.2 Calculation of the Horizontal Seismic Action — 077
  - 4.3.3 Calculation of the Vertical Seismic Action — 082
- 4.4 Seismic Check — 084
  - 4.4.1 Seismic Check for the Load-Bearing Capacity of Structural Members — 085
  - 4.4.2 Seismic Check for Deformation — 086

Chapter 5 Seismic Design of RC Structures

- 5.1 Introduction — 092
- 5.2 Earthquake Damage and Analysis — 093
  - 5.2.1 Damage to Frame Members — 093
  - 5.2.2 Damage to Shear Walls — 093
  - 5.2.3 Damage to In-Filled Walls — 094
  - 5.2.4 Other Damages — 094
- 5.3 Structural Systems and Seismic Grading — 094
  - 5.3.1 Selection of Structural System — 094
  - 5.3.2 Structural Configuration — 095
  - 5.3.3 Seismic Joints — 097
  - 5.3.4 Seismic Grading — 098
- 5.4 Seismic Design of RC Frames — 100
  - 5.4.1 General Design Scheme — 100
  - 5.4.2 Seismic Response Calculation — 100
  - 5.4.3 Internal Force under Seismic Action — 101
  - 5.4.4 Internal Force under Vertical Loads — 103
  - 5.4.5 Loads Combination — 109
  - 5.4.6 Cross-Section Design — 110
  - 5.4.7 Lateral Displacement Check — 111

Chapter 6 Seismic Design of Masonry Building Structures

- 6.1 Introduction — 114
  - 6.1.1 Masonry Structures — 114
  - 6.1.2 Ductile Design — 114
- 6.2 Structural Configuration and Systems — 114
  - 6.2.1 Structural Configuration — 114
  - 6.2.2 Principle of Horizontal Seismic Action Distribution — 116
- 6.3 Basic Seismic Design for Multi-Storey Masonry Buildings — 117
  - 6.3.1 Equivalent Wall Stiffness — 117
  - 6.3.2 Distribution of Shear Force on Floors — 119
  - 6.3.3 Seismic Shear Bearing Capacity of Walls — 120
- 6.4 Basic Seismic Design of Multi-Storey Masonry Buildings with the First One/Two Frame-Supported Storeys — 121
- 6.5 Limitations and Reasons — 122
- 6.6 Construction Measures — 124
  - 6.6.1 Construction Measures for Multi-Storey Brick Masonry Buildings — 124
  - 6.6.2 Constructional Measures for Multi-Storey Block Masonry Buildings — 128
  - 6.6.3 Constructional Measures for Multi-Storey Masonry Buildings with Frame-Shear Walls at Lower Storeys — 130

Chapter 7 Seismic Design of Steel Building Structures

- 7.1 Introduction — 136
  - 7.1.1 Earthquake Disaster Characteristics of Steel Structure Buildings — 136
  - 7.1.2 Seismic Fortification Goal of Steel Structures — 139
- 7.2 Steel Structures for Middle and High-Rise Building — 140
  - 7.2.1 Seismic Conceptual Design — 140

	7.2.2	Computation of Earthquake Action	148
	7.2.3	Seismic Checking for Members and Sections	151
	7.2.4	Detailing Requirements for Seismic Design of Members	156
7.3		Seismic Design of Steel Structures of Single-Storey Factories	161
	7.3.1	Seismic Design Concepts	161
	7.3.2	Calculation of Earthquake Action	164
	7.3.3	Checking Members and Details	165
7.4		An Example of Seismic Design	168

Chapter 8 Seismic Design of Non-Structural Elements

8.1		Introduction	178
8.2		Key Points for Seismic Design	179
	8.2.1	Influence of Non-Structural Elements on Main Structure Calculation	179
	8.2.2	Seismic Calculation of Non-Structural Elements	180
	8.2.3	Requirements for the Equivalent Lateral Force Method	181
	8.2.4	Calculation Requirement for the Floor Response Spectrum Method	181
	8.2.5	Combination of Seismic Action Effect	183
8.3		Basic Seismic Measures of Architectural Non-Structural Elements	183
8.4		Basic Seismic Measures for Mechanical and Electrical Equipment Support	186
8.5		Simplified Seismic Analysis Method	187
	8.5.1	Time-Historey Analysis Method for Calculating the Seismic Response of Equipment	188
	8.5.2	Practical Calculation Method for the Seismic Response of Equipment on the Floor	188
	8.5.3	Verification of the Practical Seismic Response Calculation Method	189

Chapter 9 Introduction to Seismic Isolation and Energy Dissipation for Building Structures

9.1		Introduction	192
	9.1.1	Seismic Isolation	192
	9.1.2	Seismic Energy Dissipation	192
9.2		Seismic Isolation for Building Structures	193
	9.2.1	Types of Multi-Layer Rubber Bearings	194
	9.2.2	Summary	195
	9.2.3	Application	195
9.3		Seismic Energy Dissipation for Building Structures	196
	9.3.1	Characteristics of Energy Dissipation Technology	196
	9.3.2	Energy Dissipation Devices	198
	9.3.3	Applications	200

References

Chapter 1
Earthquakes and Ground Motions

1.1 Causes and Types of Earthquakes

Earthquakes are a natural phenomenon that manifests itself as ground shaking resulting from the sudden release of energy stored inside the earth. Approximately, 5,000,000 earthquakes occur each year around the world. Only 1% of these earthquakes can be felt, known as felt earthquakes because most earthquakes occur at great depths or the energy released is relatively small. Intense earthquakes that cause disasters occur less frequently, about 10 per year. Shaking and ground rupture are the principal effects of earthquakes, primarily resulting in severe damage to the ground and buildings and loss of life. Furthermore, earthquakes can cause fires, floods, landslides, avalanches and tsunamis, which can be disastrous to human beings.

1.1.1 Internal Structure of Earth

To better understand the cause and development of an earthquake, the structure of the earth is briefly presented.

The earth is an oblate spheroid with an average radius of 6371 km, an equatorial radius of 6378 km and a polar radius of 6357 km. The earth is generally divided into three layers based on chemical or physical properties: crust, mantle and core. The earth section and crust section are shown in Fig. 1-1.

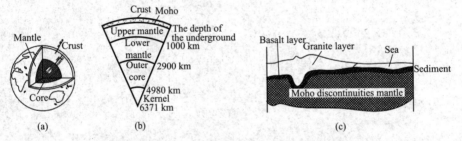

Fig. 1-1 The earth section and crust section
(a) The earth section; (b) The laminar structure; (c) The crust section

1. Earth's Crust

The crust, or the lithosphere, which is the outermost layer, is comprised of various rocks. The crust is separated from the mantle by the Mohorovičić discontinuity (or Moho). The thickness of the crust varies from 16 to 40 km on land, and is thicker at higher altitudes, reaching 70 km on the Chinese Tibet Plateau and in the Tianshan District. It is considerably thinner beneath oceans, averaging from 10 to 16 km, and the thinnest part of the crust is approximately 5 km. The continental crust is primarily composed of granitic rocks overlying the basaltic rock, whereas the oceanic crust is composed of basalt. The vast majority of earthquakes occur in this layer.

2. Earth's Mantle

The mantle extends to a depth of 2895 km and represents approximately 5/6 of the volume of the earth. The mantle is assumed to be composed of peridotite, which is an ultramafic rock. The mantle structure consists of (1) a homogeneous layer of pyroxene and olivine in the upper mantle reaching a depth of about 400 km and (2) a homogeneous layer in the lower mantle made of magnesium, iron oxide and quartz. The core-mantle boundary is known as the Gutenberg discontinuity. No earthquakes have been recorded in the lower mantle. Since shear waves (transverse seismic waves) propagate through the mantle, it has been suggested that the mantle should be solid. The pressure at the top of the mantle is approximately 24 GPa (2.45×10^5 kg/cm^2) while deeper at the core-mantle boundary it reaches 140 GPa (1.42×10^6 kg/cm^2).

3. Earth's Core

The core extends from the Gutenberg discontinuity to the center of the earth with a radius of 3500 km, which represents 17% of the earth's volume and 33% of its mass, approximately. This concludes that denser materials exist in the core of the earth. The core is composed of two parts: the inner core with a radius of 1370 km and the outer core (1300 km < radius < 3500 km). The core is essentially consisted of iron but also some nickel. The outer core is believed to be liquid because no seismic waves have been observed in this region, and the inner core behaves as a solid.

Both density and temperature increase with increasing depth in the earth's interior. The densities are between 9×10^3 and 12×10^3 kg/m^3 in the outer core while 12×10^3 and 13×10^3 kg/m^3 in the inner core.

1.1.2 Causes of Earthquakes

Earthquakes can be the result of many natural and human-induced phenomena. Natural phenomena include tectonic plates shift, volcanic activities, geological faults and rockslides. Human-induced phenomena include blasting, explosive tests, demolitions, groundwater extraction, geothermal operations and rock stress changes induced by the filling of large man-made reservoirs. However, the vast majority of damaging earthquakes originate from or are adjacent to the boundaries of crustal tectonic plates due to the relative deformations at the boundaries.

A coherent global explanation for most earthquakes can be given in terms of plate tectonics, which is now considered to be the most reliable theory. The relative motion between these tectonic plates leads to increasing stress and stored strain energy in the volume around the fault surface. This pattern continues until the stress reaches a sufficient level, causing fracture propagation along a fault plane, which releases the stored energy. The energy is released as a combination of radiated elastic strain seismic waves, frictional heating of the fault surface and cracking of the rock, which causes an earthquake. A fault that emerges at the surface of the earth because of an earthquake is known as an earthquake fault. Faults are causes rather than the results of earthquakes. Tectonic earthquakes often occur at existing faults due to stress concentration and low-strength rock.

Faults are classified in terms of the direction and nature of the relative displacement of the earth at the fault plane. There are three main types of earthquake faults: normal, reverse (thrust) and strike-slip. Normal and reverse faults are examples of dip-slip, where the displacement along the fault is in the direction of the dip, and the movement on them involves a vertical component. A normal fault (Fig. 1-2a) is one in which the rock above the inclined fault surface moves downward relative to the underlying crust. Furthermore, faults with nearly vertical slip are included in this category. A reverse fault (Fig. 1-2b) is one in which the crust above the inclined fault surface moves upward relative to the block below the fault. Strike-slip faults (Fig. 1-2c), which is also called transcurrent fault or lateral fault, involve lateral displacements of the rock, i.e., parallel to the strike. A number of earthquakes are due to movement on the faults that have components of all three fault types.

1.1.3 Types of Earthquakes

An earthquake is a sudden release of energy in the earth caused by various forces, e.g., a rock burst, collapse or volcanic eruption, resulting in ground motion. The place where the earthquake originates in the interior of the earth is known as the "hypocenter" or "focus". Although the hypocenter has a certain range, it is always considered a point in seismology because

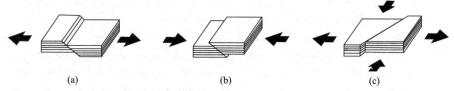

Fig. 1-2 Types of earthquake faults
(a) Normal fault; (b) Reverse fault; (c) Lateral fault

it is relatively small when compared to other aspects of seismology. The projection of the hypocenter on the earth's surface is known as the "epicenter" (Fig. 1-3). For obvious reasons, the destruction caused by the earthquake at the epicenter will always be maximum and the intensity of the destruction decreases as one moves away from this point. The distance between the hypocenter and the epicenter is known as the focal depth. Earthquakes occurring at a depth of less than 60 km are classified as "shallow-focus" earthquakes, whereas those with a focal depth between 60 and 300 km are commonly termed "mid-focus" or "intermediate-depth" earthquakes. Deep-focus earthquakes occur at considerably greater depths, i. e., exceeding 300 km.

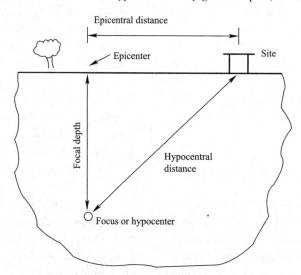

Fig. 1-3 Anatomy of an earthquake

The majority of earthquakes are shallow-focus earthquakes, with a focal depth ranging from 5 to 20 km, whereas intermediate-depth earthquakes occur less frequently, and deep-focus earthquakes rarely happen. Generally, for the same earthquake magnitude, an earthquake with a smaller focal depth results in a larger extent of damage over a small area, whereas an earthquake with a greater focal depth results in less damage but over a larger area. An earthquake with a focal depth greater than 100 km usually does not cause damage at the surface.

Earthquakes can also be classified as foreshocks, main shocks and aftershocks. In some earthquake sequences, foreshocks are smaller earthquakes that precede the larger earthquake event, which can be used to predict the main shock. Aftershocks are hundreds and even thousands of smaller earthquakes that follow the largest earthquake (main shock), some of which may have relatively high intensity and may lead to more shifts and collapse to damaged buildings, resulting in more casualties.

1.2 Seismic Waves and Propagation

The energy released by an earthquake from the focus propagates in all directions in the form of seismic waves. Mainly there are two types of seismic waves: body waves and surface waves. These seismic waves are similar to waves in air or water in several important aspects. The characteristics of these two kinds of waves are different.

1.2.1 Body Waves

Body waves originate from the rupture zone and travel through the earth's inner layer. They include P waves (also known as primary, push-pull waves, longitudinal waves, compressional waves, etc.) and S waves (also called secondary waves, shear waves, transverse waves, etc.).

The P waves shake the medium that they travel through in parallel to the direction in which they propagate. The motion of the P wave exhibits the same properties as sound waves. It pushes (compression) and pulls (tension) the ground alternately as shown in Fig. 1-4(a). P waves have short periods and small amplitudes, with relatively low damage potential. P waves are the fastest seismic waves and the first to reach the seismograph. They can travel through solids and liquids or gas, such as rocks, volcanic magma, water and atmosphere.

The S waves shake the medium that they travel through perpendicular to the direction of propagation, producing an up-and-down and side-to-side motion that shakes the ground vertically and horizontally at right angles as indicated in Fig. 1-4(b). The S wave motion introduces shear stresses in the rock and can cause major damage to structures. S waves are the second seismic waves to reach the seismograph during an earthquake. They can travel only through solids, and their amplitude is significantly reduced in liquefied soils.

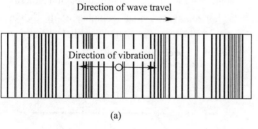

Fig. 1-4 Diagrammatical sketch of the propagation of body waves
(a) P waves; (b) S waves

The propagation velocities of the P and S waves depend on the density and elastic properties of the ground they pass through, which can be expressed as follows:

$$V_P = \sqrt{\frac{E(1-\nu)}{\rho(1+\nu)(1-2\nu)}} = \sqrt{\frac{\lambda+2G}{\rho}} \quad (1\text{-}1)$$

$$V_S = \sqrt{\frac{E}{2\rho(1+\nu)}} = \sqrt{\frac{G}{\rho}} \quad (1\text{-}2)$$

where,

V_P——the velocity of the P wave;
V_S——the velocity of the S wave;
E——Young's modulus;
ν——the Poisson ratio;
ρ——the mass density of the medium;
G——the shear modulus;
λ——the Lame constant.

$$\lambda = \frac{\nu E}{(1+\nu)(1-2\nu)}$$

Using Eqs. (1-1) and (1-2), the ratio of

V_P to V_S can be obtained as follows:

$$\frac{V_P}{V_S} = \sqrt{\frac{2(1-\nu)}{1-2\nu}} \quad (\text{usually} > 1)$$

(1-3)

For general materials, $V_P > V_S$. Thus, during earthquakes, P waves travel faster than S waves. For example, if the Poisson ratio for the earth's body is assumed to be 0.25, then $V_P = \sqrt{3} V_S$.

The time difference between arrival of P and S waves at an observation point estimates the distance to the earthquake epicenter. The time difference is easily obtained using a strong-motion earthquake accelerogram. The accelerograph is triggered and starts recording when the P wave arrives, which produces a small acceleration. The acceleration sharply increases when the S wave arrives.

Based on Eqs. (1-1)—(1-3), the parameters of the material can be obtained. For example, using the measured V_P and V_S, the Poisson ratio can be calculated using Eq. (1-3). If the mass density ρ and (E, G), (ν, λ) or (V_P, V_S) are known, then the other two sets of parameters can be obtained. These parameters are extremely important in earthquake engineering research and applications.

1.2.2 Surface Waves

Surface waves are seismic waves that travel mainly along the surface of the earth rather than deep in the interior. They are generally considered to be generated by the reflection of body waves with the stratum interface. Surface waves travel more slowly than body waves. Therefore, at the time of occurrence of the earthquake, P waves arrive first, then S waves and surface waves are the last to reach the seismograph. The most intense ground motion occurs when the S waves or surface waves arrive. They can travel through solids and liquids. There are two types of surface waves known as Love (noted as "L or LQ waves") and Rayleigh (noted as "R or LR waves") waves.

Rayleigh waves move in two directions at once up-down and push-pull motion along the direction in which the wave is moving. These motions generate an elliptic motion, as shown in Fig. 1-5(a). Rayleigh waves appear far from the epicenter and its amplitude decreases exponentially as the distance from the surface increases.

Love waves move the ground side to side in a horizontal plane perpendicular to the direction in which the wave is traveling, as shown in Fig. 1-5(b). Due to the coupling of the horizontal vibration with wave propagation, there is a horizontal torsional component in a Love wave, which is one of the essential characteristics of a Love wave.

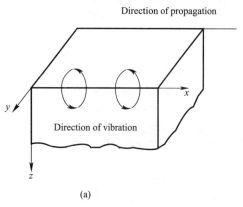

Fig. 1-5 Surface waves propagation
(a) Rayleigh waves; (b) Love waves

1.2.3 Earthquake Records

The seismic waves released from the hypocenter cause ground motion. The ground motion can be measured in terms of acceleration, velocity or displacement of the particles on the ground in terms of time. The instruments that measure the ground displacements are called seismographs. The record obtained from a seismograph is called a seismogram. As accelerations are the causal phenomena for the forces that damage structures (Newton's second law of motion: Force = mass × acceleration), engineers are more interested in acceleration measurement. The instruments that measure the ground accelerations are called accelerometers and the record obtained is called an accelerogram. The characteristics of the seismic acceleration time history, which are widely used in earthquake engineering applications, are listed below.

1. Peak Ground Acceleration

The peak ground acceleration (PGA) primarily influences the vibration amplitudes and is equal to the maximum ground acceleration and the largest increase in velocity recorded during an earthquake. It has been extensively used to scale earthquake design spectra and acceleration time histories.

2. Contents of the Frequency Spectrum

The frequency content of ground motions can be examined by transforming the motion from a time domain to a frequency domain using the Fourier transform (FT). The power spectral density (PSD), response spectrum and Fourier spectrum are commonly used to characterize the frequency content. The ground motion has a long period component for soft soil and a short period component for stiff soil, which is extremely useful in the seismic design of structures to avoid resonance.

3. Duration

Buildings sometimes collapse due to a single large earthquake pulse, but mostly they collapse for long repeated vibrations and deformations. The duration is thus used to check the collapse-resistant performance of buildings.

1.3 Earthquake Magnitude and Intensity

The severity of an earthquake can be described by both its magnitude and intensity.

1.3.1 Earthquake Magnitude

Magnitude is one of the fundamental and important parameters of an earthquake. It is a number (depicted in Arabic numerals) that defines the size of an earthquake by indirectly measuring the amount of energy released by the geological fracture. There are different approaches to measure the magnitude of an earthquake: Richter (or local) magnitude M_L, surface wave magnitude M_S and body wave magnitude M_b.

Richter magnitude, M_L: In 1935, C. F. Richter defined the magnitude of a local earthquake as the logarithm to the base 10 of the maximum amplitude (displacement) in microns (1×10^{-6} m) recorded on a Wood-Anderson seismograph located at a distance of 100 km from the earthquake epicenter.

$$M_L = \log A \qquad (1\text{-}4)$$

where,

A——the maximum amplitude of the seismic record in μm;

M_L——the magnitude of the earthquake.

However, a standard seismograph is not always set at a point 100 km from the epicenter, thus, the following formula can be used:

$$M_L = \log A - \log A_0 \qquad (1\text{-}5)$$

where,

A_0——the maximum amplitude of a standard seismic record at the same epicenter distance.

The local magnitude M_L and the magnitude of an earthquake can be identified as follows:

$$M_L = \log A_\mu + R(\Delta) \quad (1\text{-}6)$$

where,

A_μ——the maximum amplitude of the local seismic record in μm, which is the arithmetic mean of the two horizontal components; the amplitudes of the two components do not have to be measured at the same time.

$R(\Delta)$ can be determined as follows:

$$R(\Delta) = \log V_0(T) - \log A_0 - 3 \quad (1\text{-}7)$$

where,

$V_0(\cdot T)$——the amplification in the maximum amplitude period recorded by the standard seismograph;

A_0——the same as that in Eq.(1-5).

Surface wave magnitude M_S used in China can be obtained as follows:

$$M_S = \log(A/T) + \sigma(\Delta) + C \quad (1\text{-}8)$$

where,

A——the maximum displacement amplitude of the surface waves in μm, which is vector sum of the two horizontal components;

T——the period corresponding to the amplitude A;

$\sigma(\Delta)$——a function of the starting calculation;

C——the calibration value of the station.

For deep-focus earthquakes, only small surface waves can be recorded; hence, Gutenberg suggests using the body wave magnitude.

Body wave magnitude M_b used in China can be obtained as follows:

$$M_b = \log(A/T) + Q + S \quad (1\text{-}9)$$

where,

A——the maximum amplitude of the body wave in μm, in which the horizontal component is the vector sum of the two horizontal components;

T——the period corresponding to the amplitude A;

Q——the function of the starting calculation;

S——the calibration value of the station.

Theoretically, M_L, M_S and M_b should be equivalent for an earthquake. However, system deviations are observed for different magnitudes. The empirical formulas are as follows:

$$M_S = 1.13 M_L - 1.08 \quad (1\text{-}10)$$
$$M_b = 0.63 M_S + 2.5 \quad (1\text{-}11)$$

The Chinese earthquake department has selected M_S as the reported earthquake magnitude for convenience.

The energy released by an earthquake E can be approximately estimated by converting the surface wave magnitude M_S to energy using the following formula:

$$\log E = 11.8 + 1.5 M_S \quad (1\text{-}12)$$

where, E is measured in joules. Thus, the ratio of energy released by two earthquakes differing by 1 in magnitude is equal to 31.6. The ratio differing by 2 in magnitude is 1000 for earthquakes.

As a general guideline, earthquakes with a magnitude less than 2, called minor earthquakes, can barely be felt by people. Earthquakes with a magnitude between 2 and 4, known as felt earthquakes, can be felt by humans. Earthquakes with a magnitude greater than 5, known as destructive earthquakes, can cause damage. Finally, earthquakes with a magnitude greater than 7 are known as strong earthquakes.

1.3.2 Earthquake Intensity

Earthquake intensity is a number (expressed in Roman numerals) describing the degree of destruction caused by an earthquake. In other words, it indicates the local effects and the potential for damage produced by an earthquake on the earth's surface affecting humans, animals, structures and natural spaces.

Several different intensity scales have been proposed around the world over the past century to assess the effects of earthquakes. The most widely used intensity scale is the Modified Mercalli Intensity (MMI) scale, a 12-grade proposed by Wood and Neumann in 1931 and is currently used in North America and several other countries. Another intensity scale is the

Medvedev-Sponheuer-Karnik (MSK) Scale, a 12-grade scale first proposed in 1964, generally employed in Central and Eastern Europe and several other countries. The European Macroseismic Scale (EMS) is a 12-grade scale, a developed version of the MMI scale adopted since 1998 in Europe. The 7-grade scale of the Japanese Meteorological Agency (JMA) has been used since 1949 in Japan. In China, the China Seismic Intensity Scale (CSIS) is used to measure seismic intensity, a 12-grade intensity scale similar to the MMI. First proposed in 1980 by the China Earthquake Administration (CEA), it was later revised and adopted as a national standard. The last standard revision was set in 2020 (Table 1-1).

Chinese Seismic Intensity Scale (2020)　　　　Table 1-1

Intensity	Sensed by people on the ground	Extent of building damage			Other damage	Horizontal ground motion	
		Type	Damage	Mean damage index		Peak acceleration (m/s^2)	Peak speed (m/s)
I (1)	Insensible	—	—	—	—	1.80×10^{-2} ($<2.57 \times 10^{-2}$)	1.21×10^{-3} ($<1.77 \times 10^{-3}$)
II (2)	Sensible only by very few indoor people who are at rest	—	—	—	—	3.69×10^{-2} (2.58×10^{-2} —5.28×10^{-2})	2.59×10^{-3} (1.78×10^{-3} —3.81×10^{-3})
III (3)	Only sensible by a few indoor people who are at rest	—	Doors and windows rattle slightly	—	Slight swing of suspended objects	7.57×10^{-2} (5.29×10^{-2} —1.08×10^{-1})	5.58×10^{-3} (3.82×10^{-3} —8.19×10^{-3})
IV (4)	Sensible by most people inside the building and some people outside. Some sleeping people wake up	—	Doors, windows and vessels rattle	—	Clear swinging of hanging objects, rattling of pots	1.55×10^{-2} (1.09×10^{-1} —2.22×10^{-1})	1.20×10^{-2} (8.20×10^{-3} —1.76×10^{-2})

Continued Table

Intensity	Sensed by people on the ground	Extent of building damage			Other damage	Horizontal ground motion	
		Type	Damage	Mean damage index		Peak acceleration (m/s^2)	Peak speed (m/s)
V(5)	Generally, sensible by almost all people indoors and most people outdoors, and most of the sleeping people wake up	—	Doors, windows, roofs and roof trusses vibrate and rattle; dust fall, small cracks in plaster, some roof tiles and bricks falling from chimneys	—	Suspended objects swing with great amplitude. Objects in unstable positions shakes or overturns	3.19×10^{-1} (2.23×10^{-1} — 4.56×10^{-1})	2.59×10^{-2} (1.77×10^{-2} — 3.80×10^{-2})
VI(6)	Most people can no longer stand safely, and some run outside in fear	A1	Slight damage; most in good condition	0.02—0.17	Crevices in river banks and soft ground, sand volcanoes from water-saturated layers of sand, cracks in individually standing chimneys	6.53×10^{-1} (4.57×10^{-1} — 9.36×10^{-1})	5.57×10^{-2} (3.81×10^{-2} — 8.17×10^{-2})
		A2	Slight damage; majority in good condition	0.01—0.13			
		B	Very slight damage; most in good condition	≤0.11			
		C	Very slight damage; majority in good condition	≤0.06			
		D	Very slight damage; majority in good condition	≤0.04			

Continued Table

Intensity	Sensed by people on the ground	Extent of building damage			Other damage	Horizontal ground motion	
		Type	Damage	Mean damage index		Peak acceleration (m/s^2)	Peak speed (m/s)
VII(7)	The majority walk outside in fear, which is also noticeable to motorcyclists and people in moving cars	A1	Slight damage	0.15—0.44	Collapse of river banks; frequent burst of sand and water from saturated sand layers; many crevices in soft ground, moderate destruction of most of the individual chimneys	1.35(9.37× 10^{-1}—1.94)	1.20×10^{-1} (8.18×10^{-2} —1.76× 10^{-1})
		A2	Slight damage, most in good condition	0.11—0.31			
		B	Very slight damage, most in good condition	0.09—0.27			
		C	Very slight damage, most in good condition	0.05—0.18			
		D	Very slight damage, majority in good condition	0.04—0.16			
VIII(8)	Most people sway a lot and walking is difficult	A1	Mostly moderate damage	0.42—0.62	Crevices even in hard and dry ground. Severe destruction of most of the individual chimneys, the breaking of tree tops, death of people and cattle from building destruction	2.79(1.95 —4.01)	2.58×10^{-1} (1.77×10^{-1} —3.78× 10^{-1})
		A2	Mostly moderate damage	0.29—0.46			
		B	Mostly slight damage	0.25—0.50			
		C	Mostly slight damage	0.16—0.35			
		D	Mostly slight damage	0.14—0.27			

Continued Table

Intensity	Sensed by people on the ground	Extent of building damage			Other damage	Horizontal ground motion	
		Type	Damage	Mean damage index		Peak acceleration (m/s^2)	Peak speed (m/s)
IX(9)	Moving people fall	A1	Extensive severe damage	0.60—0.90	Many crevices even in hard and dry ground, possibly crevices and shifts in solid rock. Landslides and basic movements are common. The collapse of many individual chimneys, the breaking of tree tops, death of people and cattle from building destruction	5.77(4.02—8.30)	5.55×10^{-1} (3.79×10^{-1}—8.14×10^{-1})
		A2	Extensive severe damage	0.44—0.62			
		B	Few collapses and moderate damage	0.48—0.69			
		C	Few collapses and slight damage	0.33—0.54			
		D	Few collapses and slight damage	0.25—0.48			
X(10)	Cyclists fall, people in unstable state fall down, feeling of being thrown up	A1	Many collapses	0.88—1.00	Breakage of solid rock and formation of earthquake crevices, the collapse of bridge arches founded in rock, damage to the foundation or the collapse of most of the individually standing chimneys	1.19×10 (8.31—1.72×10)	1.19(8.15×10^{-1}—1.75)
		A2	Many collapses	0.60—0.88			
		B	Several collapses	0.67—0.91			
		C	Extensive severe damage	0.52—0.84			
		D	Extensive severe damage	0.46—0.84			

Continued Table

Intensity	Sensed by people on the ground	Extent of building damage			Other damage	Horizontal ground motion	
		Type	Damage	Mean damage index		Peak acceleration (m/s^2)	Peak speed (m/s)
XI(11)		A1	The vast majority collapse	1.00	Long-lasting earthquake crevices, extensive landslides	2.47×10 (1.73×10 —3.55×10)	2.57(1.76 —3.77)
		A2		0.86—1.00			
		B		0.90—1.00			
		C		0.84—1.00			
		D		0.78—1.00			
XII(12)			Almost completely collapse	1.00	Nearly damaged totally. Drastic changes in the landscape, mountains and rivers	>3.55×10	>3.77

Notes:

1. Buildings in Table 1-1 include the following three types:

   Type A: Old-style houses built with wooden frames, soil, stone and brick walls;

   Type B: Single-storey or multi-storey masonry structures without seismic resistance;

   Type C: Single-storey or multi-storey masonry structures with seismic resistance for intensity 7.

2. The peak ground acceleration and peak ground velocity are reference values and the values in parentheses denote the corresponding range.

Note for the intensity ratings:

(1) Intensities 1 to 5 depend primarily on human feeling; intensities 6 to 10 depend primarily on building damage and landscape effects—here human perception is only used as additional information, and intensities 11 and 12 depend primarily on changes in the landscape and the building damage.

(2) General buildings include wooden buildings, antique buildings built with soil, stone and brick walls and new masonry buildings. The extent of earthquake damage and the earthquake damage index in Table 1-1 can be increased or decreased for high-quality or low-quality structures based on specific conditions.

(3) An earthquake damage index of 0 indicates that a building is in perfect condition, whereas a 1 indicates that the building is completely destroyed. The average earthquake damage index is the sum of each earthquake damage index multiplied by the corresponding destruction rate.

(4) The severity of the damage caused by the earthquake can be classified as follows:

(a) Essentially intact: the structural and non-structural components are intact, cracks occur, and the structure can be used without repairs. The earthquake damage index range is $0.00 \leqslant d < 0.10$.

(b) Slight damage: visible cracks in individual structural components, and obvious cracks in non-structural components, and can be used without repair or minor repairs. The earthquake damage index range is $0.10 \leqslant d < 0.30$.

(c) Moderate damage: most structural components have slight cracks, and some have obvious cracks. Some non-structural components are seriously damaged and can be used after general repairs. The earthquake damage index range is $0.30 \leqslant d < 0.55$.

(d) Severe damage: most of structural components are seriously damaged, and non-structural components partially collapse, making it difficult to repair. The earthquake damage index range is $0.55 \leqslant d < 0.85$.

(e) Destroyed: most of structural components are seriously damaged, the structures of the houses are on the verge of collapse or has been destroyed, and there is no possibility of repair. The earthquake damage index range is $0.85 \leqslant d \leqslant 1.00$.

(5) Temporary supplements according to specific conditions.

(6) Villages and towns can assess the magnitude of their own regions, and the area is suitable to approximately 1 km².

(7) Notes about qualifiers: "very few" means < 10%; "few" means 10%—50%; "much" means 50%—70%; "most" means 70%—90%; and "commonly" means > 80%.

1.3.3 Relationship between Magnitude and Intensity

The severity of an earthquake can be described by both its magnitude and its intensity. These two terms refer to different but related aspects of an earthquake. Ideally, any given earthquake can be described by a single magnitude but several intensities because the earthquake effects vary with the circumstances, such as the distance from the epicenter and the local soil conditions. For moderate and shallow-focus earthquakes, the relationship between the magnitude and the epicenter intensity is listed in Table 1-2.

Relationship between the magnitude and the epicenter intensity   Table 1-2

Magnitude (M)	2	3	4	5	6	7	8	More than 8
Intensity (I_0)	I (1) and II (2)	III (3)	IV (4) and V (5)	VI (6) and VII (7)	VIII (8)	IX (9) and X (10)	XI (11)	XI (12)

1.4 Earthquake Characteristics and Hazards in China

1.4.1 Earthquake Activity and Distribution in China

China is located between the Circum-Pacific Seismic Zone and the Eurasian Seismic Zone, which are two of the most active seismic zones in the world. Therefore, China experiences severe disasters and extensive loss of life from earthquakes. Based on historical earthquake records, Taiwan has the most earthquakes, followed by Xinjiang and Tibet, and the southeast, northwest, north and southeast coastal areas of China have comparatively more destructive earthquakes.

According to the ***Seismic Fortification Intensity Zonation Map of China*** (2015), the areas and the percentage of the total land area are listed in Table 1-3 for the different intensity categories.

Area and percentage of the total area for the different intensity categories

Table 1-3

Intensity	<Ⅵ(6)	Ⅵ(6)	Ⅶ(7)	Ⅷ(8)	≥Ⅸ(9)	Total
Total area (10^4 km^2)	201	361	320	68	9.5	959.5
Percentage (%)	21	38	33	7	1	100

As indicated in Table 1-3, 79% of the total area, i.e., the area of intensity ≥6 has to be designed to be seismic resistant based on the *Code for Seismic Design of Buildings* (GB 50011—2010) Version 2016 and *General Code for Seismic Precaution of Buildings and Municipal Engineering* (GB 55002—2021), and 8% of the total area, i.e., the area of intensity ≥8, has high seismicity.

There are 34 high-intensity zones (seismic intensity≥9), and most of them are located in western China, including 24 on the Qinghai-Tibet Plateau and surrounding areas, 6 in Xinjiang, 2 in North China and 2 in Taiwan. There are 24 low-intensity zones (seismic intensity ≤ 7), and they are primarily located in the Huanan region, the northern region of Inner Mongolia, the northeast region and the northwest region.

The intensities of the provincial capital cities and other important cities can be given in Table 1-4.

The intensities of different cities in China

Table 1-4

City	Intensity	City	Intensity	City	Intensity
Beijing	Ⅷ(8)	Shanghai	Ⅶ(7)	Tianjin	Ⅶ(7)
Guangzhou	Ⅶ(7)	Shenyang	Ⅶ(7)	Wuhan	Ⅵ(6)
Nanjing	Ⅶ(7)	Chengdu	Ⅶ(7)	Fuzhou	Ⅶ(7)
Jinan	Ⅵ(6)	Zhengzhou	Ⅶ(7)	Kunming	Ⅷ(8)
Changsha	Ⅵ(6)	Harbin	Ⅵ(6)	Changchun	Ⅶ(7)
Taiyuan	Ⅷ(8)	Shijiazhuang	Ⅶ(7)	Hohhot	Ⅷ(8)
Hangzhou	Ⅵ(6)	Hefei	Ⅶ(7)	Nanchang	Ⅵ(6)
Nanning	Ⅵ(6)	Guiyang	Ⅵ(6)	Lhasa	Ⅷ(8)
Xi'an	Ⅷ(8)	Lanzhou	Ⅷ(8)	Xining	Ⅶ(7)
Yinchuan	Ⅷ(8)	Haikou	Ⅷ(8)	Urumqi	Ⅷ(8)
Taipei	Ⅷ(8)	Chongqing	Ⅵ(6)	Hong Kong	Ⅶ(7)
Macao	Ⅶ(7)	Shenzhen	Ⅶ(7)	Zhuhai	Ⅶ(7)
Xiamen	Ⅶ(7)	Dalian	Ⅶ(7)	Qinhuangdao	Ⅶ(7)
Yantai	Ⅶ(7)	Qingdao	Ⅵ(6)	Lianyungang	Ⅶ(7)
Nantong	Ⅵ(6)	Ningbo	Ⅵ(6)	Wenzhou	Ⅵ(6)
Zhanjiang	Ⅶ(7)	Beihai	Ⅵ(6)	Shantou	Ⅵ(6)

1.4.2 Characteristics of Earthquake Activity in China

1. Wide Distribution

Based on the historical records, most of the provinces in China have suffered earthquakes with magnitudes larger than 6 Ms. Because the seismic activity is widely distributed, the epicenters are dispersed, and technologies are limited; therefore, recording the specific locations where earthquakes occur as well as resisting the earthquakes is difficult.

2. Shallow-Focus and High-Intensity Earthquakes

Most earthquakes occur in mainland China, and the majority of which are shallow-focus earthquakes with a focal-depth of 20—30 km. Shallow-focus earthquakes cause severe damage to buildings and facilities. Deep-focus earthquakes with a focal-depth larger than 300 km or approximately 400—500 km have occurred in Jixi, Yanji, Tibet and Xinjiang. Over the past 80 years, 1/10 of the earthquakes with magnitudes larger than 7 Ms in the world occurred in China, and 2/10—3/10 of the global energy released by earthquakes occurred in China.

3. Several Large Cities in the Seismic Region and the Lack of Earthquake-Resistant Buildings

There are 450 cities in China; 335 cities are located in the seismic region, and half of them are of intensities of at least 7. Among the 28 megacities with a population of more than a million, 85.7% are located in the seismic region. Moreover, 80% of the large cities with a population of 0.2—1 million are located in the seismic region. Several important cities, such as Beijing, Kunming, Taiyuan, Hohhot, Lhasa, Xi'an, Lanzhou, Urumqi, Yinchuan, Haikou and Taipei, are located in a high seismic intensity region with intensities larger than 8.

Before the 1970s, construction projects did not consider seismic precautionary measures. Therefore, these buildings and facilities cannot resist earthquakes. Most old buildings in cities, buildings in the countryside, soil and stone buildings and buildings with cavity walls in southern China have a considerably lower seismic resistance. The loss of life and property from each earthquake is primarily attributed to the low seismic resistance of buildings and facilities.

4. Long Return Period of Strong Earthquakes

The return period of strong earthquakes is hundreds of years in China. Thus, predicting the scope of earthquake activities and seismic intensities is difficult. There are several cases in which a region with intensity 6 suffered strong earthquakes with an intensity greater than 6. For example, in eastern China, where the population, cities and industries are concentrated, after an 8 Ms earthquake occurred in Quanzhou, Fujian Province, in 1604, an 8.5 Ms earthquake occurred in Yancheng, Shandong Province, in 1668, an 8 Ms earthquake occurred in the Sanhe, Pinggu area of the Hebei province in 1679, and an 8 Ms earthquake occurred in Linfen, Shanxi Province, in 1695. After that, no 8 Ms earthquake has occurred in approximately 280 years. 3 earthquakes with magnitudes larger than 7.5 Ms has occurred in the Hebei Province, and the time intervals between the earthquakes are 151 years and 146 years. And 3 earthquakes with magnitudes larger than 7.5 Ms has occurred in the Shanxi Province, and the time intervals are 791 years and 392 years. Furthermore, 2 strong earthquakes has occurred in Yancheng and Heze, Shandong Province, and the time interval between these 2 earthquakes is 269 years. Because of the long return period of strong earthquakes, it's easy to overlook threats and forget the painful lessons learned from the dangers caused by earthquakes. Therefore, a lack of knowledge regarding the threats and risk preparedness caused by earthquakes will most likely lead to greater disasters.

1.4.3 Earthquake Hazards in China

China suffers from some of the world's most severe earthquake risks. The earthquakes in China cause severe damage and loss of lives and financial impact. The damage caused by earthquakes primarily includes the following items.

1. Ground Damage

Generally, ground damage caused by earthquakes includes surface cracking, sand and water blasting and landslide.

One type of surface cracking is a rupture caused by the dislocation of the underground fault due to a strong earthquake. Another type is crossed cracks produced where the soil is soft, such as an original watercourse, embankment shore and steep slope. The latter type has different shapes and is smaller than the first type.

Generally, sand and water blasting occurs in coastal areas or places with high groundwater levels. The strong shaking squeezes and pressurizes the aquifer. Groundwater with sand blasts from the surface cracks or areas with soft soil.

Landslide always occurs in steep mountainous regions. Earthquakes can produce slope instability, which leads to landslide.

2. Damage to Building Structures

The damage and loss of life and money due to earthquakes mainly result from the damage to building structures.

The seismic damage inflicted on building structures can be divided into five states: no damage, slight damage, moderate damage, severe damage and collapse. The criteria for the classification are as follows:

(1) No damage: no damage in structural members; slight damage in several non-structural members.

(2) Slight damage: slight cracks in several structural members; moderate damage in non-structural members.

(3) Moderate damage: slight cracks in most structural members; severe damage in several non-structural members.

(4) Severe damage: severe damage in structural members; partial collapse.

(5) Collapse: most structural members collapse.

The multi-storey masonry building damage due to destructive earthquakes in the 1960s in China is listed in Table 1-5. The table indicates that masonry structures have always been destroyed in different damage states; however, collapse occurs primarily when the seismic intensity exceeds 9.

Multi-storey masonry building damage in China   Table 1-5

Damage states	6		7		8		9		10	
	Number	%	Number	%	Number	%	Number	%	Number	%
No damage	230	45.9	250	40.8	141	37.2	7	1.6	4	0.3
Slight damage	212	42.3	231	37.7	74	19.5	35	7.8	30	2.5
Moderate damage	56	11.2	75	12.2	94	24.8	138	30.7	66	5.6
Severe damage	3	0.6	54	8.8	69	18.2	169	37.5	154	13.0
Collapse	—	—	3	0.5	1	0.3	101	22.4	933	78.6
Total	501	100	613	100	379	100	450	100	1187	100

The damage inflicted upon single-storey reinforced concrete (RC) factory structures due to earthquakes in China is listed in Table 1-6. Generally, no damage occurs to structural members for intensity 7, several structural members are damaged for intensity 8, and there is severe damage to the structural members for intensity 9.

Damage to single-storey RC factory structures in China   Table 1-6

Damage states	7		8		9		10	
	Number	%	Number	%	Number	%	Number	%
No damage	3	15.8	24	13.7	—	—	—	—
Slight damage	11	57.9	46	26.3	1	10.0	3	6.7
Moderate damage	3	15.8	59	33.7	2	20.0	15	33.3
Severe damage	2	10.5	38	21.7	7	70.0	11	24.4
Collapse	—	—	8	4.6	—	—	16	35.6
Total	19	100		100	10	100	45	100

RC frame structures have better seismic performance. The damage always occurs in infilled walls, weak stories, column ends and corner columns. The inner-frame structures suffer considerably more damage. Table 1-7 represents the damage to inner-frame structures.

Damage to inner-frame structures in China   Table 1-7

Damage states	7		8		9		10	
	Number	%	Number	%	Number	%	Number	%
No damage	1	3.4	3	25.0	3	10.0	—	—
Slight damage	10	34.5	1	8.3	3	10.0	3	6.0
Moderate damage	14	48.3	3	25.0	6	20.0	1	2.0
Severe damage	4	13.8	2	16.7	14	46.7	5	10.0
Collapse	—	—	3	25.0	4	13.3	42	82.0
Total	29	100	12	100	30	100	51	100

3. Secondary Disaster

An earthquake may spawn a secondary disaster, such as fire, flood, tsunami, poisonous gas or air pollution, which increases the impact. In 1906, the fire that followed the San Francisco earthquake nearly destroyed the city, in which 28,000 buildings were destroyed. However, the losses caused by the fire were 4 times that caused by the earthquake. In 1964, during the Niigata earthquake in Japan, a tsunami hit Niigata City approximately 15 minutes after the earthquake caused flooding damage and it persisted in some areas for up to a month. In 1970, the debris flow of rock, ice and snow induced by the Peru earthquake (also known as the Great Peruvian earthquake) destroyed and buried villages, and 25,000 people died. In 1960, earthquakes occurred in Chile, and 22 hours later, a tsunami approached Honshu and Hokkaido, located 17,000 km from Chile. The tsunami destroyed the harbor, wharf and structures around the coast. In 1993, a gas leak and fire induced by the Hokkaido earthquake threatened 224 people. The tsunami caused by the Indonesia earthquake in December 2004 resulted in tremendous casualties and property losses. In May 2012, the Wenchuan earthquake occurred, and landslides, mudslides and lakes caused by the earthquake resulted in severe economic losses and secondary hazards.

1.5 Seismic Fortification for Building Structures

1.5.1 Seismic Fortification Objectives

The objectives of seismic fortification are to prioritize the prevention of earthquake disasters and to minimize damage and loss of buildings, life and economy.

In general, the seismic fortification objectives determined by Chinese *Cocle for Seismic Design of Buildings* (GB 50011—2010) Version 2016 are "No damage under minor earthquakes, repairable under moderate earthquakes and no collapse under major earthquakes", which are often referred to as the three levels of fortification.

1. The first level: No damage under minor earthquakes

When buildings designed based on the code are subjected to frequent earthquakes of an intensity of lower than the fortification intensity of the region, they will not or will only be slightly damaged and should continue to be serviceable without repair.

2. The second level: Repairable under moderate earthquakes

When buildings are subjected to earthquakes equal to the fortification intensity of the region, they may be damaged but should still be serviceable after ordinary repair or without repair.

3. The third level: No collapse under major earthquakes

When buildings are subjected to the influence of expected rare earthquakes with an intensity higher than the fortification intensity of the region, they should not collapse or suffer damage that would endanger human lives.

1.5.2 Seismic Design Methods

The Chinese seismic fortification objectives are achieved by different design methods and requirements according to different levels, which is called a "two-stage-and-three-level seismic design method". It is detailed as follows:

1. The first level

By the elastic analysis, the carrying capacity of the structure is checked under the fundamental combination of effects of seismic action of minor earthquake and other loads, and the elastic seismic deformation is checked under the action of minor earthquake.

2. The second level

The objective of this level is realized mainly by seismic conceptual design and seismic measures or detailing.

3. The third level

Under rare earthquakes, the seismic design should be carried out in accordance with the collapse prevention requirements. For the brittle structure, seismic measures should be taken to strengthen it. For ductile structures, especially those liable to collapse during an earthquake, the elastic-plastic deformation is checked under the action.

The basic seismic intensity of a region denotes the maximum seismic intensity that would most likely be encountered in the region in a certain future period under ordinary site conditions, i.e., the intensity specified in the current "Chinese Seismic Intensity Zoning Map". Protection intensity is the intensity approved by the state authority as the basis for the protection of structures against earthquakes in a certain region.

Chapter 2
Site, Subsoil and Foundation

2.1 Site

Structures come in different shapes, forms and sizes. However, all structures have foundations through which the superstructure interfaces with the underlying soil or rock. In a seismic environment, the loads imposed on a foundation of a structure under seismic excitation can greatly exceed the static vertical loads, and may even produce uplift. Besides, there will be horizontal forces and potential moments at the foundation level. Therefore, the potential for ground failures may exist. It is obvious that if the ground fails beneath a structure, it could be severely or completely damaged.

2.1.1 Selection of Construction Site

1. Site Classification

In the selection of a construction site, the features that are favorable or unfavorable to the seismic design of buildings can be described as follows:

(1) Favorable to earthquake resistance: steady bedrock, stiff soil, dense and homogeneous medium-stiff soil in a wide-open area.

(2) Unfavorable to earthquake resistance: soft soil, liquefiable soil, band-shaped protruding ridge, high isolated hills, non-rocky steep slopes, river banks and edges of slopes, soil strata having heterogeneous distribution in-plane (such as abandoned and filled riverbeds, fracture zones, hidden swamps, streams, gullies and pits as well as subsoil with partial excavation and backfilling).

(3) Hazardous to earthquake resistance: places where landslide, avalanche subsidence, the formation of cracks and mudflows, rock flow liable to occur during an earthquake, active fault zone.

(4) Ordinary to earthquake resistance: other ground sections.

During the selection of a construction site, a comprehensive evaluation should be performed based on the project requirements, the seismicity of the region and the geotechnical and geological data available for the site. The areas favorable to earthquake resistance should be selected, whereas unfavorable areas should be avoided. If unfavorable areas are unavoidable, appropriate seismic measures should be taken. Type A, B and C buildings must not be built in hazardous seismic regions.

2. Avoidance Measures for Causative Faults

If a causative fault exists at a site, the impact of the causative fault on the planned projects should be evaluated. For conditions satisfying one of the following requirements, the impact of the causative fault dislocation on the surface structures may be neglected.

(1) The precautionary seismic intensity is less than 8;

(2) No Holocene active faults;

(3) For intensities 8 and 9, the soil layer coverage for the hidden fault exceeds 60 m and 90 m, respectively.

If the situation does not comply with the above provisions, the primary fault zone should be avoided when selecting a building site. Also, the avoidance distance should be at least the minimum avoidance distance of the causative fault, which are specified in Table 2-1. In the case of Type C and D buildings with less than 3 storeys that are required to be built within the scope of the avoidance distance, the seismic measures of one intensity higher should be taken, the integrity of the foundation and superstructure should be improved, and the building should not cross over any fault trace.

Minimum avoidance distance of the causative fault (m)　　Table 2-1

Intensity	Precautionary category of building			
	A	B	C	D
Ⅷ(8)	Special study required	200	100	—
Ⅸ(9)	Special study required	400	200	—

2.1.2 Construction Site Classification

1. Seismic Influence to Construction Site

Local soil conditions have an extremely important role for the response of structures to earthquakes. The soil and rock at a site have specific characteristics that can significantly amplify the incoming earthquake motions traveling from the earthquake source. The importance of local site conditions was recognized in the 1960s due to the influence of the ground motions on mid-height buildings during the Caracas and Venezuela earthquakes. For buildings with approximately the same height and similar construction, buildings founded on soft soils suffered more damage than similar buildings founded on rock soils. These observations were subsequently confirmed during the Mexico City earthquake in 1985, where the ground motions were amplified in the deep soft lake bed deposits that lay under the city; these ground motions had a long period and adversely affected several high-rise buildings.

Low-frequency accelerations can be amplified at soil sites, especially those containing soft layers. Conversely, high-frequency accelerations can be amplified in rock. In addition to the peak rock acceleration, several factors, including soil softness and layering depth, play a role in the degree of amplification. One important factor is the impedance contrast between the soil and the underlying rock. Earthquake records for soft to medium clay sites indicate that the soil or rock amplification factors for long-period spectral accelerations can be significantly large. Furthermore, the largest amplification often occurs when the dominant period of an earthquake is close to the natural period of the soil deposit. During the Mexico City earthquake in 1985, the maximum rock acceleration was amplified by a factor of 4 due to the soft clay deposits. An inspection of the records obtained for certain soft clay sites during the Loma Prieta earthquake in 1989 indicates a maximum amplification of 3 to 6 times the long-period spectral amplitudes. Thus, in addition to peak rock acceleration, layer thickness and soil softness play a key role in the degree of amplification.

2. Classification of Site Soil

The site soil can be classified according to the shear wave velocity V_s of the soil layer, which is indicated in Table 2-2.

Classification of site soil
Table 2-2

Type of soil	Shear wave velocity of the soil layer (m/s)
Rock	$V_s > 800$
Stiff soil	$500 < V_s \leqslant 800$
Medium-stiff site soil	$250 < V_s \leqslant 500$
Medium-soft site soil	$150 < V_s \leqslant 250$
Soft site soil	$V_s \leqslant 150$

If no measured data of shear wave velocity for the building of Type C to D is available, the soil can be classified according to Table 2-3, where f_{ak} is the characteristic value of load-bearing capacity of subsoil in "kPa".

When the site contains several types of soil layers, the equivalent shear wave velocity should be used to determine the soil type. The equivalent shear wave velocity v_{se} is determined as follows:

$$v_{se} = d_0/t \qquad (2\text{-}1)$$

$$t = \sum_{i=1}^{n} (d_i/v_{si}) \qquad (2\text{-}2)$$

Classification of the soil — Table 2-3

Type of soil	Geotechnical description	Shear wave velocity of the soil layer (m/s)
Rock	Stiff and hard rock	$V_s > 800$
Stiff soil	Stable rock, dense gravel	$500 < V_s \leqslant 800$
Medium-stiff soil	Dense, medium dense or slightly dense gravel; coarse or medium sand; cohesive soil and silt with $f_{ak} > 150$	$250 < V_s \leqslant 500$
Medium-soft soil	Slightly dense gravel; coarse or medium-sized sand; fine and silty sand (excluding the loose sand), cohesive soil and silt with $f_{ak} \leqslant 150$; filled soil with $f_{ak} \geqslant 130$	$150 < V_s \leqslant 250$
Soft soil	Muck and muddy soil, loose sand, new alluvial sediment of cohesive soil and silt, backfill with $f_{ak} < 130$	$V_s \leqslant 150$

where,

v_{se} —— the equivalent shear wave velocity of the soil layer;

d_0 —— the depth of the soil (m), equals to 20 m but no longer than the thickness of the overlying layer;

d_i —— the thickness of the layer (m);

t —— the travel time of the shear wave from the ground level to the calculation depth;

v_{si} —— the shear wave velocity in layer i (m/s);

n —— the number of soil layers.

The overlaying thickness at a building site in the formula above should be determined based on the following requirements:

(1) Generally, this thickness should be determined based on the distance from the ground surface to the top surface of a soil layer at which the shear wave velocity is more than 500 m/s and the shear wave velocity of the soil layers under this layer is of at least 500 m/s.

(2) If the shear wave velocity of a soil layer at least 5 m below the ground surface is more than 2.5 times that of the soil layers above and the shear wave velocity of this soil layer and those under it is of at least 400 m/s, the cover layer thickness may be determined based on the distance from the ground surface to the top surface of this soil layer.

2.1.3 Construction Site Categories

Construction sites shall be classified into five categories according to the equivalent shear wave velocity and the overlaying thickness of the site. If a reliable shear wave velocity and the overlaying thickness are available and their values are close to the division between site classes listed in Table 2-4, the character period for the earthquake action calculation can be determined via interpolation.

The thickness of overlaying layers at various sites (m) — Table 2-4

Equivalent shear wave velocity (m/s)	Construction site categories				
	I_0	I_1	II	III	IV
$V_s > 800$	0	—	—	—	—
$500 < V_s \leqslant 800$	—	0	—	—	—
$250 < V_s \leqslant 500$	—	<5	$\geqslant 5$	—	—
$150 < V_s \leqslant 250$	—	<3	3—50	>50	—
$V_s \leqslant 150$	—	<3	3—15	15—80	>80

[Example 2-1]

The soil profile for a site is presented in Table 2-5. To determine the category of the construction site.

Data of soil exploration at site Table 2-5

Bottom depth of soil layers (m)	Thickness of soil layers (m)	Soil profile name	Shear wave velocity (m/s)
9.5	9.5	Sand	170
37.8	28.3	Very soft clay	138
48.6	10.8	Sand	240
60.1	11.5	Very soft silty clay	200
68.0	7.9	Fine sand	330
86.5	18.5	Gravelly sand	550

[Solution]:

(1) Determine the overlying thickness of the site

The shear wave velocity of the soil layer beneath 68 m is more than 500 m/s, so the overlay thickness of the site is equal to 68 m (>20 m).

(2) Determine the type of soil within the range of 20 m below ground level

$$t = \sum_{i=1}^{n} (d_i/v_{si}) = 9.5/170 + 10.5/138 = 0.132 \text{ s}$$

The equivalent shear wave $v_{se} = d_0/t = 20 \div 0.132 = 151.6$ m/s.

The equivalent shear wave velocity is 150 m/s $<V_S<$ 250 m/s, so the type of soil is medium-soft soil.

(3) Determine the construction site category

The equivalent shear wave velocity is 150 m/s $<V_S<$ 250 m/s and the overlying thickness of the site is larger than 50 m; hence, the soil belongs to the Type III group based on Table 2-4.

2.2 Seismic Checking for Subsoil and Foundation

2.2.1 Earthquake Resistance of Natural Subsoil

For the following types of buildings, the bearing capacity of natural subsoil and foundation do not need be checked for earthquake resistance.

1) Structures, of which seismic checking need not be conducted for their superstructures specified in the code.

(Most buildings are constructed in the site with the intensity fortification not larger than 6, not including tall buildings.)

2) The following structures, of which no soft clay layer exists in their force-bearing subsoil.

(1) Ordinary single-storey factory buildings, and single-storey spacious buildings;

(2) Masonry buildings;

(3) Multi-storey framed civil buildings (frame and frame-shear wall structures that not exceeding 24 m in height and not exceeding 8 storieys);

(4) Multi-storey framed and frame-shear wall factory buildings (with equivalent foundation loading to that of multi-storey framed civil buildings).

Note: Soft clay refers to the soil with a characteristic value of load-bearing capacity of subsoil (f_{ak}) is less than 80 kPa, 100 kPa and 120 kPa for intensities 7, 8 and 9, respectively.

2.2.2 Earthquake-Resistant Checking for Natural Subsoil and Foundation

With the exception of the buildings listed above, the bearing capacity of natural subsoil needs to be checked for earthquake resistance. Using the following formula:

$$f_{aE} = \xi_a f_a \qquad (2\text{-}3)$$

where,

f_{aE}——the adjusted characteristic value of seismic bearing capacity of the subsoil;

ξ_a——the adjusting coefficients for seismic bearing capacity of subsoil, and taken from Table 2-6;

f_a——the characteristic value of load-bearing capacity of subsoil with the modification of width and depth; to check the vertical bearing capacity of a natural base under an earthquake, the mean pressure on the base of the foundation and the maximum pressure at the edge of the foundation shall comply with the following formulas:

$$p \leqslant f_{aE} \qquad (2\text{-}4)$$
$$p_{max} \leqslant 1.2 f_{aE} \qquad (2\text{-}5)$$

where,

p——the mean design value of pressure of combined seismic action on the foundation base;

p_{max}——the maximum design value of pressure of combined seismic action at the edge of the foundation base.

Note: No zero-stress area of the foundation should occur for high-rise buildings with an aspect ratio greater than 4. For other buildings, the zero-stress area between the foundation base and the subsoil shall not be more than 15% of the total area of the foundation base. The zero-stress area for structures such as the chimney should comply with the corresponding specifications of the current national standard.

Adjustment coefficients for the seismic bearing capacity of base   Table 2-6

Name and characteristics of the rock and/or soil	ξ_a
Rock; dense detritus; dense gravel, coarse and medium-sized sand; cohesive soil and silt with $f_{ak} \geqslant 300$ kPa	1.5
Moderately dense and slightly dense detritus; moderately dense and slightly dense gravel; coarse and medium-sized sand; dense and moderately dense fine and mealy sand; cohesive soil and silt with 150 kPa $\leqslant f_{ak} < 300$ kPa; hard loess	1.3
Slightly dense fine and mealy sands; cohesive soil and silt with 100 kPa $\leqslant f_{ak} < 150$ kPa; plastic loess	1.1
Silt; silty soil; loose sand; fill; newly alluvial sediment deposit of loess or flowing mucky loess	1.0

2.2.3 Earthquake-Resistant Checking for Pile Foundation

For the following buildings, in which pile foundations have low pile caps and mainly support the vertical load, there is no liquefied soil layers underground, no silt or silty soil surrounding the pile cap or no backfill with characteristic value of the load-bearing capacity not greater than 100 kPa, the seismic bearing capacity of the pile foundation needs not to be checked. A seismic bearing capacity check is not needed for the following scenarios:

(1) The following buildings located in regions of intensities 7 and 8:

(a) Ordinary single-storey factory buildings and single-storey large space buildings;

(b) Ordinary frame structure buildings that do not exceed 8 storeys and 24 m in height;

(c) Multi-storey frame structure factory buildings and multi-storey buildings with concrete walls, for which the foundation load is equivalent to that specified in Item (b).

(2) Buildings that use pile foundations among those specified in Item 1) and Item 2) in Section 2.2.1 of this chapter.

The seismic check on a low-cap pile foundation in non-liquefied soil should meet the following requirements:

(1) The characteristic value of both the vertical and horizontal seismic bearing capacities of an individual pile should increase by 25% compared with those of a non-seismic design.

(2) When the backfill soil surrounding the pile cap is tamped so that the dry density at least meets the requirements specified in the current national standard *Code for Design of Building Foundation* (GB 50007—2011), the fill and the pile work together to endure the horizontal seismic actions. However, the frictional forces between the bottom surface of the cap and the foundation soil should not be considered.

2.3 Soil Liquefaction

2.3.1 Cause and Damage of Soil Liquefaction

Liquefaction is a term used to describe a range of phenomena in which the strength and stiffness of soil deposits are reduced due to the generation of pore water pressure. Liquefaction results from seismic shaking with sufficient intensity and long duration, causing the settlement of structures, landslides, failures of earth dams and other types of hazards. Liquefaction most commonly occurs in loose saturated sands. During strong earthquake shaking, a loose saturated sand deposit will tend to compact and, thus, have a decrease in volume. If this deposit cannot drain rapidly, there will be an increase in the pore water pressure, which even makes the effective stress of soil particles equal to zero. Since the shear strength of the soil is directly proportional to the effective stress, the sand will not have any shear strength and is now in a liquefied state, and liquefaction occurs. "Sand boils" existing at the ground surface during an earthquake provide evidence of liquefaction. Because the capacity of the soil is related to its strength, liquefaction poses a serious hazard to constructed structures.

Liquefaction can have a significant effect on buildings supported on upper soils if the consequences of liquefaction are not considered. Large movements may occur because the lateral support is temporarily missing, causing high-bending moments in the pile. Liquefaction occurs in numerous earthquakes. Liquefaction can affect a wide variety of civil structures and facilities by altering the ground motion and the development of permanent deformations. In fact, the extent to which structures are directly or indirectly affected by liquefaction-caused failures depends on the extent of liquefaction. If the soil becomes liquefied and loses its shear strength, ground failures may result. Generally, the ground failures caused by liquefaction can be classified into three categories.

1. Lateral Spreading

Lateral spreading is the movement of the surface soil layers in a direction parallel to the ground surface, which occurs when there is a loss of shear strength in a subsurface layer due to liquefaction. Lateral spreading can produce significant and damaging lateral displacements of the ground surface. Furthermore, lateral spreading typically occurs on extremely gentle slopes. If there is differential lateral spreading under a structure, sufficient tensile stresses could develop in the structure, forcing the structure to tear apart. Hence, lateral spreading can have a catastrophic impact on long, linear buried utilities or, as some may prefer, "lifelines".

2. Flow Failure

Among the most significant risks caused by earthquakes are the flow failures induced by the liquefaction of sandy slopes and their effects on adjacent structures. This phenomenon occurs when large zones of soil liquefy or when blocks of non-liquefied soil flow over a layer of liquefied soil. Flow slides can develop, where the slopes are generally steep. On flat ground, ground oscillations can occur because deep liquefaction decouples the overlying surface layers from the underlying liquefied soil. The decoupling causes the upper surface layers to oscillate with occasionally large displacements or visible ground waves.

3. Loss of Bearing Capacity

Liquefaction can also lead to a marked reduction in the load-bearing capacity of the soil. During liquefaction, the shear strength of the soil can be reduced to a residual strength which is a function of the density of the soil. If the residual strength is lower than the shear stress required to maintain static equilibrium, the bearing capacity can decrease. Additionally, buried structures have been observed to "float" out of the ground. An evaluation of the potential for such instabilities requires an estimate of the residual strength of the liquefied soil.

2.3.2 Discrimination of Soil Liquefaction

Not all soils are susceptible to liquefaction; therefore, the first step in performing a liquefaction risk evaluation is to determine the liquefaction potential. The liquefaction potential can be evaluated using historical, geologic, compositional and state criteria. This section introduces the Chinese code for determining the liquefaction potential.

A two-step discrimination method is adopted to determine the liquefaction potential in the *Code for Seismic Design of Buildings* (GB 50011—2010) Version 2016. This method includes preliminary discrimination and the standard penetration test method. A block diagram of the first step in determining the liquefaction potential is illustrated in Fig. 2-1.

1. Preliminarily Discrimination

When the intensity is 6, discrimination of the liquefaction potential of saturated soil and the adoption of measures to prevent liquefaction usually should not be considered, except for Type B buildings which are sensitive to the settlement caused by liquefaction, measures for intensity 7 may be used. Furthermore, when the intensity is 7 to 9, for Type B buildings, discrimination of the liquefaction potential and the adoption of relevant measures may be considered using those specified for the original intensity.

If there are saturated sandy soil and saturated silt underground, a liquefaction evaluation should be performed, except for buildings with intensity 6; for a base with a liquefied soil layer, corresponding measures should be considered based on the precautionary category of the building and the extent of liquefaction along with certain specific conditions.

If one of the following conditions is satisfied, saturated sand or saturated silt may be preliminarily discriminated as non-liquefiable soil, or effects of liquefaction need not be considered:

(1) If the geological period of the soil is Pleistocene of the Quaternary (Q3) period or earlier, the soil may be considered non-liquefiable when the intensity is 7 or 8;

(2) If the clay particle content (particle diameter less than 0.005 mm) of silt is not less than 10%, 13% and 16%, when the intensity is 7, 8 and 9 respectively, the soil may be considered non-liquefiable;

(3) For buildings resting on natural subsoil, the effects of liquefaction need not be considered when the thickness of the non-liquefiable overlaying layer and the depth of underground water level comply with one of the following conditions:

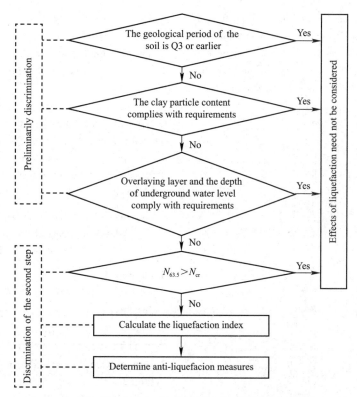

Fig. 2-1 Flow chart for determining the liquefaction potential

$$d_u > d_c + d_b - 2 \quad (2\text{-}6)$$
$$d_w > d_c + d_b - 3 \quad (2\text{-}7)$$
$$d_u + d_w > 1.5 d_c + 2 d_b - 4.5 \quad (2\text{-}8)$$

where,

d_w——the depth of the groundwater level (m), for which the mean annual highest level during the service of the building should be used or the annual highest level in recent years may also be used;

d_u——the thickness of the non-liquefiable overlaying layer (m), in which the thickness of silt and silty soil layer should be deduced;

d_b——the buried depth of the foundation (m), if it is not more than 2 m, then 2 m shall be used;

d_c——the characteristic depth of liquefaction-potential soil (m), values in Table 2-7 may be used.

2. Standard Penetration Tests (SPT)

The standard penetration test (SPT) is a dynamic in-situ penetration survey designed to provide information on the geotechnical engineering properties of soil. If the test is considered in the preliminary discrimination that further discrimination of the liquefaction potential is necessary, the standard penetration tests shall be performed. The tests aim to determine the value of the standard penetration tests, which gives an indication of the soil stiffness and can be empirically related to many engineering properties.

Characteristic depths of liquefaction-potential soil (m) Table 2-7

Type of saturated soil	Intensity		
	Ⅶ(7)	Ⅷ(8)	Ⅸ(9)
Silt	6	7	8
Sand	7	8	9

Fig. 2-2 illustrates the equipment used for standard penetration test. The test is carried out as follows. First, the penetrator is driven to a depth of nearly 15 mm above the testing soil layer. It is then hammered into the ground by a drop hammer weighing 63.5 kg, falling from a height of 76 cm. The number of hammer blows required to push the penetrator 30 mm deeper into the testing soil layer is recorded as the measured value of the standard penetration test.

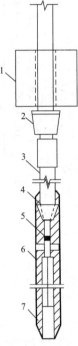

Fig. 2-2 Equipment of the standard penetration test
1—Drop hammer with through tang; 2—Pallet; 3—Penetration sounding pole; 4—Penetrator; 5—Outflow hole; 6—Body of penetrator; 7—Jacket of penetrator

Within a depth range of 20 m under the ground level, the following formulas shall be satisfied:

$$N_{63.5} < N_{cr} \quad (2\text{-}9)$$
$$N_{cr} = N_0 \beta [\ln(0.6 d_s + 1.5) - 0.1 d_w] \sqrt{3/\rho_c} \quad (2\text{-}10)$$

where,

$N_{63.5}$——the measured value of the standard penetration resistance in terms of the number of blow counts for saturated soil (correction for the length of the rod not yet considered);

N_{cr}——the critical value of the standard penetration resistance in terms of the number of blow counts for liquefaction discrimination;

N_0——the reference value of the standard penetration resistance in terms of the number of blow counts for liquefaction discrimination, which should be obtained from Table 2-8;

d_s——the depth of the standard penetration for saturated soil (m);

d_w——the depth of the underground water level (m);

ρ_c——the percentage of the clay particle content, when it is less than 3% or when the soil is sand, the value of 3% shall be used;

β—— the coefficient is taken 0.80, 0.95, and 1.05 for seismic groups 1, 2 and 3, respectively.

Reference values of the standard penetration resistance in terms of the number of blow counts   Table 2-8

Peak ground acceleration (g)	0.10	0.15	0.20	0.30	0.40
N_0	7	10	12	16	19

3. Liquefaction Index and Category of Liquefaction

For the subsoil with liquefaction-potential soil layers, the depth and thickness of each layer shall be investigated and the liquefaction index shall be calculated using the following formula:

$$I_{lE} = \sum_{i=1}^{n} \left(1 - \frac{N_i}{N_{cri}}\right) d_i W_i \quad (2\text{-}11)$$

where,

I_{lE}——the liquefaction index;

n——the total number of standard penetration test points in each bore within the depth of discrimination under the ground surface;

N_i, N_{cri}—— the measured value and critical value of the standard penetration resistance in terms of the number of blow count at the i-th point, respectively; when the measured value exceeds the critical value, the latter should be used; if the liquefaction index for soil that is less than 15 m deep is calculated, then the critical value can be used for soil deeper than 15 m;

d_i——the thickness of the soil layer (m) at the i-th point, which can be taken as half of the difference in the depth between the upper and lower neighboring standard penetration test points; however, the upper point depth should be at least the depth of groundwater, and the lower point depth should not be greater than the liquefaction-potential depth;

W_i——the weighted function value of the i-th soil layer (m^{-1}) considering the effects of the layer depth of the unit soil layer thickness. When the depth of the midpoint of the layer is less than 5 m, a value of 10 should be used; when it is 20 m, a value of zero should be used; furthermore, when it is between 5 and 20 m, the value can be calculated through linear interpolation.

For the subsoil with liquefaction-potential soil layers, its category of liquefaction shall be classified based on the liquefaction index as shown in Table 2-9.

Liquefaction category   Table 2-9

Category of liquefaction	Light	Moderate	Serious
I_{lE}	$0 < I_{lE} \leqslant 6$	$6 < I_{lE} \leqslant 18$	$I_{lE} > 18$

2.3.3 Mitigation of Liquefaction Hazard

If liquefaction is identified as a hazard that could affect a structure, mitigation measures must be considered. For new construction, the available choices include:

(1) Designing for liquefaction by modifying the site soil conditions or strengthening the structure;

(2) Abandoning the project at this site;

(3) Accepting the risks by proceeding without designing for liquefaction.

Clearly, the second choice depends on whether there is an alternative site without the liquefaction problem. The third choice could bring unwanted liability exposure, uninsurability issues or even jeopardize future property values. The second and third choices could be the subject of considerable discussion, although they are beyond the scope of this study.

Anti-liquefaction measures for the subsoil can be comprehensively determined by considering the importance of the building, the category of liquefaction of the subsoil and other conditions. For flat and uniform soil layers with the liquefaction potential, the measures depicted in Table 2-10 can be selected. The untreated soil layer should not be used.

Anti-liquefaction measures   Table 2-10

Building type	Subsoil liquefaction category		
	Light	Moderate	Serious
A	(1)	(1)	(1)
B	(2) or (3)	(1) or (2) + (3)	(1)
C	(3) or (4)	(3) or (2)	(1) or (2) + (3)
D	(4)	(4)	(3) or other economic measures

Note: (1) Eliminate all layers liable to the settlement;
    (2) Eliminate partial layers liable to the settlement;
    (3) Foundation and superstructure must be treated;
    (4) No measure is taken.

The measures for eliminating all layers liable to settlement should meet the following requirements:

(1) When a pile foundation is used, the length (excluding the length of the pile tip) of the pile driven into the stable soil layer below the liquefaction depth should be calculated. For gravel, gravel sand, coarse and medium sand,

stiff cohesive soil and dense silt, the length shall not be less than 0.8 m; for other non-rocky soil, the length should not be less than 1.5 m either.

(2) When a compaction method is used for strengthening (e.g., vibration impact, vibrating compaction, sand pile compaction and strong ramming), compaction shall be carried out down to the lower margin of liquefaction potential depth, and the measured value of the standard penetration resistance of the compacted soil layer in number of blow counts shall be greater than the corresponding critical value.

(3) When the soil compaction or alteration method is used, the width of excavation out of the foundation edge shall not be less than half of the depth of excavation below the base of the foundation or not less than 2.5 m.

(4) Excavate all layers of the soil with liquefaction potential.

The measures taken to eliminate partial layers liable to settlement should comply with the following requirements:

(1) Excavation shall be carried out to a depth so that the liquefaction index shall decrease. The liquefaction index of the subsoil shall be no larger than 5. For single footings and strip footings, the depth of excavation shall not be less than the characteristic depth of liquefaction-potential soil under the base of the footing or also not less than the width of the footing, whichever is larger.

(2) In the range of the depth of excavation, the soil layers with liquefaction potential should either be excavated or strengthened by compaction so that the measured value of the standard penetration resistance of the treated soil layer (in terms of the number of blow counts) should be greater than the corresponding critical value.

The following measures can be taken to treat the foundation and the superstructure to reduce the effect of liquefaction based on comprehensive considerations:

(1) Select an appropriate buried depth for the foundation.

(2) Adjust the surface of the foundation base to reduce its eccentricity.

(3) Upgrade the integrality and rigidity of the foundation, e.g., the use of caisson or raft footing, the use of a cross-shape footing, addition of ring beams foundation or connecting beams can be considered.

(4) Decrease the load, upgrade the integrality, uniformity and symmetry of the superstructure or add settlement joints; avoid the use of a structural configuration that is vulnerable to unequal settlement.

(5) At locations where pipelines pass through the building, sufficient space shall be left beforehand for the pipelines, or flexible joints should be used.

Chapter 3
Structural Seismic Response of Single-Degree-of-Freedom and Multi-Degree-of-Freedom Systems

The vibration of structures caused by earthquakes is known as the seismic response of structures, which includes internal forces, deformations, displacements, velocity, acceleration, etc. The analysis of the seismic response of structures can be regarded as a dynamic problem; thus, the theory of structural dynamics can be applied for analyzing the seismic response of structures.

Analysis of the seismic response of structures is a necessary and important step in the design of earthquake-resilient structures and/or retrofitting of vulnerable existing structures. Since structures are usually complex systems, therefore, the simplification and idealization of the structures are essential to obtain discrete models.

The factors to be considered for the structural idealization are as follows: (1) Characteristics of building structures such as inertia (mass), elastic properties (stiffness) and energy dissipation (viscous damping); (2) Characteristics of ground motions (Section 1.2.2); (3) Purpose of seismic analysis (preliminary design), e. g. , if the analysis is only performed for the program design, the computational model can be appropriately simplified.

For example, when the structural configuration and loads are uniformly distributed along the longitudinal direction in a single-storey industrial factory, we can use one of the bays as the computing unit. Fig. 3-1 shows the computational model for this structure. If the axial deformation of each structural member (which will not be considered unless otherwise noted) is ignored, there is only one independent unknown parameter in the model, i. e. , the lateral displacement of A or B. The geometric position and deformation state can be determined when the lateral displacement is defined. In dynamic analysis, the number of degrees of freedom (DOFs) of a structural system refers to the number of independent parameters that can determine the geometric position and deformation profile of the structural system.

As depicted in Fig. 3-1, there is only one independent parameter in the structural system, i. e. , the system has only one degree of freedom, and the idealized structure is called a single degree-of-freedom (SDOF) system.

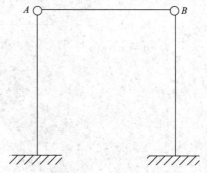

Fig. 3-1 Single-storey bent frame

Fig. 3-2 shows a sketch of a single-storey plane frame, which contains three indeterminate parameters, i. e. , the roll angles of A and B and the displacement of A or B. Thus, there are three DOFs for this system, i. e. , it is a three-degree-of-freedom (3-DOF) system.

The practical structure must be simplified and idealized to construct the computational model while performing the static analysis; then, the following question arises: what are the differences between the structural computational model under the action of static loading and its counterpart under dynamic loading?

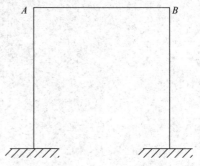

Fig. 3-2 Single-storey plane frame

For structure itself, a static computational model is the same as a dynamic computational model. However, a dynamic analysis must consider both damping and inertial forces. Therefore, dynamic analysis requires solving differential equations (discrete systems), while static analysis requires solving algebraic equations. Thus, for the same structure and computing model, a dynamic analysis is much more demanding and time-consuming compared to a static analysis. A structural dynamic computational model can be constructed based on a static computational model to make it relatively simple. Consequently, the distributed mass is typically concentrated as a lumped mass, and the degrees of freedom of small inertial forces are neglected.

For example, the single-storey plane frame (Fig. 3-2) requires up to three degrees of freedom to construct a static computational model. For the dynamic analysis, the same computational model as for the static analysis can be used, in which the mass of the columns is lumped at the two endpoints. Then, the frame can be analyzed as a three-degree-of-freedom system. To simplify the computation, the rotational inertia of the particle, i. e., two degrees of freedom, is often neglected; thus, the single-storey plane frame's computing model can be transformed into a relatively simple system, such as a SDOF system.

3.1 Free Vibration of SDOF Systems

3.1.1 Mechanical Model and the Equation of Motion

The mechanical model of the SDOF system discussed in the previous section is shown in Fig. 3-3. The physical and mechanical properties of the system are shown in Fig. 3-3 (a), and the external forces applied to the system are provided in Fig. 3-3 (b). The primary mechanical model of the linear structural system under dynamic loads consists of an inertial force, elastic force, energy dissipation mechanism or damping force and external sources of excitation or loading. Each of these internal components is assumed to be concentrated on a physical element in the SDOF system model.

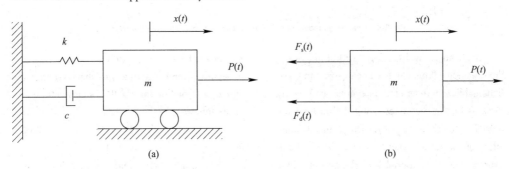

Fig. 3-3 Mechanical SDOF system model
(a) The physical and mechanical properties; (b) The external forces

At a given time t, the displacement of the mass can be denoted as x, and the corresponding velocity and acceleration are $\dot{x}$ and $\ddot{x}$, respectively. At the same time, the mass supports the elastic restoring force $F_s = -kx$, damping force $F_d = -c\dot{x}$, inertial force $F_I = -m\ddot{x}$ and applied load $P(t)$. The equation of motion can be formulated by directly expressing the equilibrium of these four forces acting on the mass applying D'Alembert's principle as follows:

$$F_s + F_d + F_I + P(t) = 0 \qquad (3-1)$$

or

$$m\ddot{x} + c\dot{x} + kx = P(t)$$

By introducing the angular frequency ω of

the mass and by expressing the viscous coefficient c as a function of the damping ratio ξ in Eq. (3-2), which represents the ratio between the viscous coefficient c and its critical value c_c, the equation of motion can be expressed as:

$$\ddot{x} + 2\xi\omega\dot{x} + \omega^2 x = a(t) \quad (3\text{-}2)$$

where,

ω——the undamped free vibration circular frequency, $\omega = \sqrt{\dfrac{k}{m}}$;

ξ——the damping ratio, $\xi = \dfrac{c}{c_c} = \dfrac{c}{2 \cdot \sqrt{k \cdot m}}$;

$a(t)$——the acceleration, $a(t) = \dfrac{P(t)}{m}$.

Eq. (3-2) is the exact equation of motion for linear SDOF structural systems.

The following discussion addresses the situation in which the ground motion is considered. It is assumed that the acceleration of the ground at a given time is $\ddot{x}_g(t)$, which is shown in Fig. 3-4. The displacement of the mass is defined as $x(t)$, whereas the corresponding velocity and acceleration are $\dot{x}(t)$ and $\ddot{x}(t)$, respectively. The absolute acceleration of the ground is $[\ddot{x}_g(t) + \ddot{x}(t)]$. Therefore, the elastic restoring force and damping force can be determined separately to be $-kx(t)$ and $-c\dot{x}$, and the inertia force is $-m[\ddot{x}_g(t) + \ddot{x}(t)]$. The equation of motion can be given as follows:

$$-m[\ddot{x}_g(t) + \ddot{x}(t)] - c\dot{x} - kx = 0 \quad (3\text{-}3)$$

$$m\ddot{x}(t) + c\dot{x}(t) + kx(t) = -m\ddot{x}_g(t) \quad (3\text{-}4)$$

or $\quad \ddot{x}(t) + 2\xi\omega\dot{x}(t) + \omega^2 x(t) = -\ddot{x}_g(t)$

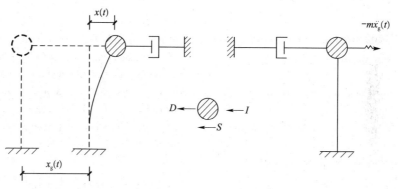

Fig. 3-4 Calculation model of a SDOF system subjected to ground motion

By comparing Eq. (3-2) with Eq. (3-4), the equation of motion for linear structural systems subjected to ground motion $\ddot{x}_g(t)$ is the same as that of the system subjected to an external load $P(t) = -m\ddot{x}_g(t)$. Thus, the dynamic effect of the ground motion can be equivalently expressed by the external load $P(t) = -m\ddot{x}_g(t)$.

3.1.2 Undamped Free Vibration of SDOF Systems

Free vibration occurs in the absence of external loads. In general, free vibration occurs when an initial velocity or displacement is imposed on a system, and the excitation of the external forces is absent. If the effects of damping are neglected, i.e., without considering the dissipative forces acting on the system, the equation of motion for SDOF systems can be given as follows:

$$m\ddot{x} + kx = 0 \quad (3\text{-}5)$$

or $\quad \ddot{x} + \omega^2 x = 0$

where $\omega^2 = k/m$, and Eq. (3-5) is a second-order linear ordinary differential equation with constant coefficients. Then, the general solution can be given as follows:

$$x(t) = B\cos\omega t + C\sin\omega t$$

The constants B and C can be determined using the initial conditions of the motion. Let the initial displacement $x(0)$ be x_0 and the initial velocity $\dot{x}(0)$ be v_0 at $t = 0$. Then the constants B and C can be determined from the ini-

tial conditions as follows:
$$B = x_0, C = v_0/\omega$$

Thus, the solution of the undamped free vibration SDOF system can be expressed as follows:
$$x(t) = x_0\cos\omega t + (v_0/\omega)\sin\omega t \quad (3\text{-}6)$$

Eq. (3-6) can be simplified to a monomial as follows:
$$x(t) = A\sin(\omega t + \psi) \quad (3\text{-}7)$$
where,
$$A = \sqrt{x_0^2 + (v_0/\omega)^2};$$
$$\tan\psi = \frac{\omega x_0}{v_0}$$

It can be easily determined that the undamped free vibration is a simple harmonic vibration, and its period can be calculated as follows:
$$T = 2\pi/\omega$$

The reciprocal of the period $f = 1/T = \omega/2\pi$ is known as the frequency. The frequency is measured in cycles per second, commonly referred to as Hz. The quantity $\omega = 2\pi f$ is known as the circular frequency or angular frequency, i.e., the number of vibrations in 2π seconds.

The frequency f or period T reflects the primary dynamic properties of the structures. Based on known conditions:
$$\omega^2 = k/m \text{ or } \omega = \sqrt{k/m}$$
Then, we can obtain the following:
$$T = 2\pi\sqrt{m/k}, f = \frac{1}{2\pi}\sqrt{k/m}$$

There is a relationship between the period of structures and both mass and stiffness. As the mass increases, the period becomes longer. Conversely, as the stiffness increases, the period becomes shorter. The natural period is a built-in attribute of these systems, i.e., it does not correlate with the external forces.

The properties of an undamped simple harmonic free vibration can be expressed as follows:
$$x(t) = A\sin(\omega t + \psi)$$
So $\ddot{x}(t) = -A\omega^2\sin(\omega t + \psi)$

Inertia force $I(t) = -m\ddot{x}(t) = mA\omega^2\sin(\omega t + \psi)$

The formulas above demonstrate that the displacement $x(t)$, acceleration $\ddot{x}(t)$ and the inertia force $I(t)$ vary following a sinusoidal function and will reach their respective maxima simultaneously.

3.1.3 Damped Free Vibration of SDOF Systems

The amplitude of the undamped free vibration is constant according to Eq. (3-7). However, motion tends to decrease with time. This reduction is associated with the loss of energy present in the system. Energy, whether kinetic or potential, is transformed into other forms of energy, such as noise, heat, etc. In a dynamic system, this loss of energy is known as damping, and the damping mechanism is extremely complex. Currently, the viscous damping theory, Coulomb damping theory, and hysteretic damping theory are commonly used to simplify the analysis. In this book, the first theory is used to simplify the calculations. Based on this theory, the damping force is proportional to the velocity and acts in the direction opposite to the motion. The equation of motion for SDOF systems formulated on the basis of the viscous damping theory and given in Eq. (3-2) is still a linear differential equation. Accordingly, the equation of motion of free vibration can be expressed as follows:
$$\ddot{x}(t) + 2\xi\omega\dot{x}(t) + \omega^2 x(t) = 0 \quad (3\text{-}8)$$

Eq. (3-8) is a second-order linear differential equation with constant coefficients. Its general solution for practical structures with a damping ratio of less than one is as follows:
$$x(t) = e^{-\xi\omega t}(B\cos\omega' t + C\sin\omega' t)$$

The parameter $\omega' = \omega\sqrt{1-\xi^2}$ is known as the circular frequency of the damped free vibration. The constants B and C can be determined based on the initial conditions as follows:
$$B = x_0, C = \frac{v_0 + \xi\omega x_0}{\omega'}$$

Thus, the solution of the damped free vibration for SDOF systems is:

$$x(t) = e^{-\xi\omega t}\left(x_0\cos\omega' t + \frac{v_0 + \omega\xi x_0}{\omega'}\sin\omega' t\right) \quad (3\text{-}9)$$

Furthermore, the equation can be simplified to a monomial as follows:
$$x(t) = Ae^{-\xi\omega t}\sin(\omega' t + \psi) \quad (3\text{-}10)$$
$$A = \sqrt{x_0^2 + \left(\frac{v_0 + \xi\omega x_0}{\omega'}\right)^2}$$

where,
$$\tan\psi = \frac{\omega' x_0}{v_0 + \xi\omega x_0}(\xi < 1)$$

Eq. (3-9) or Eq. (3-10) is the solution for the free vibration of SDOF systems.

According to Eq. (3-10), the motion of the free vibration will be damped; the amplitude will decrease. The oscillation is different from the strictly defined periodic motion.

However, the particle goes through the same equilibrium position in two successive vibrations over the period T'. Thus, the motion is known as the damped periodic free vibration. Similarly, ω' is known as the damped circular frequency, and T' is known as the damped period.

The damping ratio of practical structures is generally 0.02 to 0.05. Thus, we can assume that $\omega' = \omega\sqrt{1-\xi^2} = (0.9987\text{—}0.9998)\omega \approx \omega$, which allows us to assume $\omega' = \omega$ for practical computations.

The displacement-time history of the damped free vibration for SDOF systems is plotted in Fig. 3-5. The curve for $\xi = 0$ is the displacement-time history of the undamped free vibration.

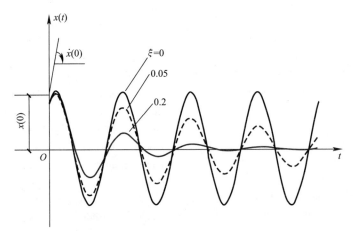

Fig. 3-5 Free vibration curves for SDOF systems

The amplitude of the undamped free vibration remains constant, while the amplitude of the damped free vibration decreases. Furthermore, the amplitude decreases faster as the damping ratio ξ increases.

[**Example 3-1**]

For a given SDOF system with mass $m = 204$ t, lateral stiffness $k = 8048.6$ kN/m and damping ratio $\xi = 0.05$, determine the properties of the free vibration of the system.

[**Solution**]:
$$\omega^2 = \frac{k}{m} = \frac{8048.6}{204.0} = 39.43$$

The circular frequency of the free vibration $\omega = 6.28$ rad/s.

The circular frequency of the damped free vibration:
$$\omega' = \omega\sqrt{1-\xi^2} = 6.28\sqrt{1-0.05^2}$$
$$= 6.27 \text{ rad/s}$$

The frequency of the free vibration $f = 2\pi/\omega = 1$ Hz.

The period of the free vibration $T = 1/f = 1$ s.

3.2 Forced Vibration of SDOF Systems under Arbitrary Loading

As mentioned previously, the dynamic effects of ground motion can be expressed by the equivalent dynamic loads $P(t) = -m\ddot{x}_g(t)$, where $\ddot{x}_g(t)$ denotes the acceleration of the ground motion at a construction site. Generally, the value of $\ddot{x}_g(t)$ is extremely variable, which is indicated in Fig. 3-6, illustrating the actual acceleration records of the measurement of the 1940 US El Centro ground motion (El Centro is very important in the history of the development of seismic design, as it is the first time that an earthquake is recorded in digital form, which could then be used for structural analysis). To deduce the dynamic response of SDOF systems under the action of an arbitrary dynamic load $P(t)$, this section will use the concept of instantaneous impulse and determine the Duhamel integral formula.

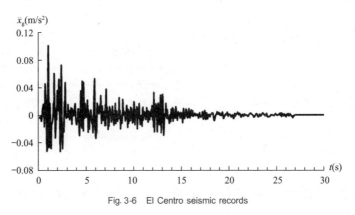

Fig. 3-6 El Centro seismic records

3.2.1 Instantaneous Impulse and its Free Vibration

Fig. 3-7 illustrates an instantaneous load $S = Pdt$. This load acts during the time interval Δt and represents a short-duration impulse on a structure, which is known as an instantaneous impulse. Under the action of the instantaneous impulse S, a static system moves at an initial velocity $v_0 = S/m$, and the initial displacement of the system is zero. Then, the system will vibrate freely at these initial conditions. Based on Eq. (3-9), i. e., the solution of SDOF systems under damped free vibration conditions, the displacement can be evaluated as follows:

$$x(t) = e^{-\xi\omega t}\frac{Pdt}{m\omega'}\sin\omega't \quad (3\text{-}11)$$

3.2.2 Dynamic Response under a General Dynamic Load-Duhamel Integration

Fig. 3-8 illustrates an arbitrary dynamic load, where the entire loading history can be treated as a series of instantaneous impulses. Based on the features of linear differential equations and the superposition principle, the total dynamic response can be evaluated by the superposition of the individual impulse responses.

Let a load $P(\tau)$ be applied at a time τ, and the impulse caused by this load in the short time interval $d\tau$ is $P(\tau)d\tau$. Based on Eq. (3-11), the response of the SDOF system produced by the impulse at time $t(t \geqslant \tau)$ is as follows:

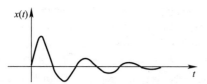

Fig. 3-7 Instantaneous impulse and free vibration

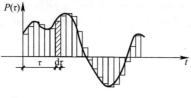

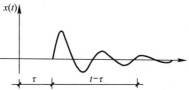

Fig. 3-8 Displacement analysis

$$\mathrm{d}x(t) = e^{-\xi\omega|t-\tau|} \cdot \frac{P(\tau)}{m\omega'}\sin\omega'(t-\tau)\mathrm{d}\tau$$

(3-12)

The reason why t in Eq. (3-11) is replaced by $t-\tau$ in Eq. (3-12) is that Eq. (3-11) is for an initial impulse at $\tau = 0$. For Eq. (3-12), to obtain the displacement caused by an impulse at a time τ, the time from τ to t is $t-\tau$.

The total displacement over the entire loading history can be obtained by summing all of the differential displacements caused by the instantaneous impulses, i. e., by integrating Eq. (3-12) as follows:

$$x(t) = \int_0^t e^{-\xi\omega|t-\tau|} \cdot \frac{P(\tau)}{m\omega'}\sin\omega'(t-\tau)\mathrm{d}\tau$$

(3-13)

Eq. (3-13), commonly known as the Duhamel integral equation, can be used to evaluate the response of SDOF systems with static initial conditions under an arbitrary dynamic load. If the system has an initial velocity or initial displacement, the corresponding resulting free vibration should be added to Eq. (3-13).

3.3 Numerical Methods for Seismic Response of SDOF Systems

3.3.1 Duhamel Integration Method

In a SDOF system subjected to a ground motion, the equivalent dynamic load is equal to $-m\ddot{x}_g(t)$. Substituting $P(t) = -m\ddot{x}_g(t)$ into Eq. (3-13), the displacement response of the system due to ground motion with an acceleration $\ddot{x}_g(t)$ can be obtained as follows:

$$x(t) = -\frac{1}{\omega'}\int_0^t \ddot{x}_g(t)e^{\omega} \cdot \sin\omega'(t-\tau)\mathrm{d}\tau$$

(3-14)

By solving the first-or second-order derivatives and finishing the integral calculation, the velocity and acceleration responses of the SDOF systems can be obtained, respectively.

Due to the extreme irregularity of the ground acceleration $\ddot{x}_g(t)$, it is generally not possible to obtain a closed form solution of Eq. (3-14). Currently, the numerical integral method (such as the Simpson's formula) is usually adopted in solving Duhamel Integral at discrete points.

3.3.2 Numerical Solutions for Equations of Motions

In addition to the method introduced above, several methods can be used to solve numerical integration of differential equations of motion, such as the mean acceleration method, linear acceleration method, Newmark-β method and Wilson-θ method. The commonly used linear acceleration method will be introduced in this section.

As depicted in Eq.(3-4), the equation of motion for a linear elastic SDOF system subjected to a ground motion acceleration is as follows:

$$\ddot{x}(t) + 2\xi\omega\dot{x}(t) + \omega^2 x(t) = -\ddot{x}_g(t)$$

To solve this differential equation using numerical methods, the time interval $[0, t]$ should be equally divided into n equivalent sections with a segment point $t_0, t_1, \cdots, t_{n-1}, t_n$, respectively, and for any point t_k, the equation of motion is:

$$\ddot{x}(t_k) + 2\xi\omega\dot{x}(t_k) + \omega^2 x(t_k) = -\ddot{x}_g(t_k) \quad (3\text{-}15)$$

The basic idea of the numerical method is to perform a recursive calculation step by step, given the initial conditions $x(t_0)$, $\dot{x}(t_0)$, $\ddot{x}(t_0)$ and the assumed recurrence relation of the segment points. Thus, the relationships between $x(t_{k-1}), \dot{x}(t_{k-1}), \ddot{x}(t_{k-1})$ and $x(t_k), \dot{x}(t_k), \ddot{x}(t_k)$ should be identified.

Suppose that the acceleration of the system varies linearly between two adjacent points, that is, in the same interval (t_{k-1}, t_k), the accelerations are:

$$\ddot{x}(t) = \ddot{x}(t_{k-1}) + \frac{t - t_{k-1}}{t_k - t_{k-1}}[\ddot{x}(t_k) - \ddot{x}(t_{k-1})]$$

Integrate the formula above on time interval $[t_{k-1}, t]$, we can obtain:

$$\int_{t-1}^{t} \ddot{x}(t)\,dt = \int_{t-1}^{t} \ddot{x}(t_{k-1})\,dt + \frac{\ddot{x}(t_k) - \ddot{x}(t_{k-1})}{t_k - t_{k-1}}$$

$$\int_{t-1}^{t}(t - t_{k-1})\,dt$$

That is $\dot{x}(t) = \dot{x}(t_{k-1}) + \ddot{x}(t_{k-1})$

$$(t - t_{k-1}) + \frac{1}{2}[\ddot{x}(t_k) - \ddot{x}(t_{k-1})]$$

$$\frac{(t - t_{k-1})^2}{t_k - t_{k-1}} \quad (3\text{-}16)$$

Integrate formula of velocity on (t_{k-1}, t), we can obtain:

$$x(t) = x(t_{k-1}) + \dot{x}(t_{k-1})(t - t_{k-1}) +$$

$$\frac{1}{2}\ddot{x}(t_{k-1})(t - t_{k-1})^2 + \frac{1}{6}\frac{\ddot{x}(t_k) - \ddot{x}(t_{k-1})}{t_k - t_{k-1}}$$

$$(t - t_{k-1})^3 \quad (3\text{-}17)$$

Substitute $t = t_k$ into the two formulas and note that:

$x_k = x(t_k), x_{k-1} = x(t_{k-1}), \dot{x}_k = \dot{x}(t_k),$
$\ddot{x}_k = \ddot{x}(t_k), \Delta t = t_k - t_{k-1}$

We can obtain:
$x_k = x(t_k), x_{k-1} = x(t_{k-1}), \dot{x}_k = \dot{x}(t_k),$
$\ddot{x}_k = \ddot{x}(t_k),$ and $\Delta t = t_k - t_{k-1}, \cdots$

Then:

$$\dot{x}_k = \dot{x}_{k-1} + \ddot{x}_{k-1} \cdot \Delta t + \frac{1}{2}(\ddot{x}_k - \ddot{x}_{k-1}) \cdot \Delta t$$

$$= \dot{x}_{k-1} + \frac{1}{2}\ddot{x}_k \cdot \Delta t + \frac{1}{2}\ddot{x}_{k-1} \cdot \Delta t \quad (3\text{-}18)$$

$$x_k = x_{k-1} + \dot{x}_{k-1} \cdot \Delta t + \frac{1}{2}\ddot{x}_{k-1} \cdot \Delta t^2$$

$$+ \frac{1}{6}(\ddot{x}_k - \ddot{x}_{k-1}) \cdot \Delta t^2$$

$$= x_{k-1} + \dot{x}_{k-1} \cdot \Delta t + \frac{1}{3}\ddot{x}_{k-1} \cdot \Delta t^2$$

$$+ \frac{1}{6}\ddot{x}_k \cdot \Delta t^2 \quad (3\text{-}19)$$

If the displacement x_{k-1}, velocity $\dot{x}_{k-1}$ and acceleration $\ddot{x}_{k-1}$ of the system at the former time t_{k-1} are known, then the displacement x_k, velocity $\dot{x}_k$ and acceleration $\ddot{x}_k$ at a time t_k can be determined by solving Eq.(3-18), Eq.(3-19) and the equation of motion Eq.(3-15).

Consider $\dot{x}_{k-1} + \frac{1}{2}\ddot{x}_{k-1} \cdot \Delta t = B_{k-1}$ in Eq.

(3-18) and $x_{k-1} + \dot{x}_{k-1} \cdot \Delta t + \frac{1}{3}\ddot{x}_{k-1} \cdot \Delta t^2 = A_{k-1}$ in Eq.(3-19).

Then, Eq.(3-18) and Eq.(3-19) can be simplified as follows:

$$\dot{x}_k = B_{k-1} + \ddot{x}_k \cdot \frac{\Delta t}{2} \quad (3\text{-}20)$$

$$x_k = A_{k-1} + \ddot{x}_k \cdot \frac{\Delta t^2}{6} \quad (3\text{-}21)$$

A formula can be obtained by substituting x_k and $\dot{x}_k$ in Eq.(3-20) and Eq.(3-21) into the equation of motion (i.e., Eq. 3-15) as follows:

$$\ddot{x}_k + 2\xi\omega B_{k-1} + 2\xi\omega\frac{\Delta t}{2}\ddot{x}_k + \omega^2 A_{k-1}$$

$$+ \omega^2 \frac{\Delta t^2}{6}\ddot{x}_k = -\ddot{x}_{gk}$$

or $\left(1 + \xi\omega\Delta t + \frac{\omega^2 \Delta t^2}{6}\right)\ddot{x}_k =$

$$-(\ddot{x}_{gk} + 2\xi\omega B_{k-1} + \omega^2 A_{k-1})$$

Consider $s = 1 + \xi\omega\Delta t + \frac{\omega^2 \Delta t^2}{6}.$

Then, $\ddot{x}_k$ is as follows:

$$\ddot{x}_k = \frac{-1}{s}(\ddot{x}_{gk} + 2\xi\omega B_{k-1} + \omega^2 A_{k-1})$$

(3-22)

As long as we know displacement, velocity and acceleration at time t_{k-1}, the response of the system at any time can be obtained by recursive calculation with the three formulas above. Initial displacement and velocity of a system are usually known to be zero, so we can achieve the seismic response of a system at any time afterward from the ground motion acceleration we've already known.

[**Example 3-2**]

There is a SDOF particle with a weight of 200 t, a lateral stiffness of $k=7200$ kN/m, and a damping ratio of $\xi=0.05$. The ground motion acceleration is provided in Table 3-1. Determine the displacements, velocities and accelerations of the mass using the linear acceleration method (the time interval is from 0 to 1.2 s, and the maximum time increment is 0.1 s).

Ground motion acceleration records　　　Table 3-1

$t(\text{s})$	0.1	0.2	0.3	0.4	0.5	0.6	0.7	0.8	0.9	1.0	1.1	1.2
$\ddot{x}_g(\text{cm/s})$	108	207	300	200	95	0	-150	-198	-120	-18	0	0

[**Solution**]

Natural frequency of vibration $\omega = \sqrt{k/m} = \sqrt{7200/200} = 6$ rad/s.

$s = 1 + \xi\omega\Delta t + \frac{\omega^2}{6}\Delta t^2 = 1 + 0.05 \times 6 \times 0.1 + \frac{6^2}{6} \times 0.1^2 = 1.09$

Based on Eq.(3-20), Eq.(3-21) and Eq.(3-22), the iteration results are listed in Table 3-2.

Iteration results　　　Table 3-2

k	$t(\text{s})$	A_k	B_k	$\ddot{x}_{gk}$	$\ddot{x}_k$	$\dot{x}_k$	x_k
0	0.0	0	0	0	0	0	0
1	0.1	0.995	-9.90	108	-99.08	-4.95	-0.17
2	0.2	-3.49	-25.08	207	-151.60	-17.49	-1.24
3	0.3	-7.46	-39.70	300	-146.16	-32.39	-3.73
4	0.4	-10.58	-31.22	200	84.75	-35.46	-7.32
5	0.5	-10.91	-3.28	95	279.06	-17.25	-10.11
6	0.6	-7.62	32.94	0	362.13	14.83	-10.31
7	0.7	-0.62	70.06	-150	371.15	51.50	-7.00
8	0.8	8.02	86.42	-198	163.56	78.24	-0.35
9	0.9	14.64	66.18	-120	-202.36	76.30	7.69
10	1.0	16.22	15.84	-18	-503.40	41.01	13.8
11	1.1	12.30	-38.60	0	-544.03	-11.38	15.31
12	1.2	4.62	-77.40	0	-388.07	-57.59	11.77

3.4　Response Spectrum for Building Design

For seismic design of building structures, the internal forces of each member of a structure under seismic actions must first be determined. Currently, there are two primary methods to determine the internal forces of structural members in buildings under seismic actions. The first method is to directly solve for the internal forces using stiffness equations according to the displacement of building structures under seismic actions, which requires a relatively accurate structure dynamic model. The second method is to determine the forces using an indirect approach by transforming the earthquake-induced dynamic problem into a static problem under static loads. Based on the acceleration responses due to ground motions, the inertia force of the structural system can be calculated and regarded as an equivalent load, which reflects the effects of the earthquake. Then, the internal forces of the structural members can be determined by static analysis, and the seismic checking calculation can be accomplished. The Chinese *Code for Seismic Design of Buildings* (GB 50011—2010) Version 2016 adopts the latter method for the design of general building structures.

3.4.1 Basic Formula for Horizontal Earthquake Action

For a SDOF system, the acceleration $\ddot{x}(t)$ of the system for a specified ground motion $\ddot{x}_g(t)$ can be obtained using the numerical methods mentioned above provided that the free frequency ω and the damping ratio ξ of the system are known. Then, the absolute acceleration of the particle can be determined as follows:

$$a = \ddot{x}_g(t) + \ddot{x}(t)$$

To obtain the maximum seismic action that a structure can endure, the maximum absolute acceleration S_a should be calculated via $S_a = |a|_{max}$; when S_a is determined, the maximum seismic action is:

$$F = mS_a \quad (3\text{-}23)$$

3.4.2 Earthquake Effect Coefficient—Response Spectrum in Chinese *Code for Seismic Design of Buildings* (GB 50011—2010) Version 2016

For the seismic design of building structures, once the structural parameters such as mass, stiffness and damping ratio are determined, the maximum absolute acceleration of the mass can be determined by inputting n proper ground motion acceleration waves selected based on the site conditions with the above numerical methods, and the seismic action can then be obtained. Although this approach is theoretically viable, it is too time-consuming for general buildings. Therefore, the Chinese *Code for Seismic Design of Buildings* (GB 50011—2010) Version 2016 introduces the concept of the earthquake effect coefficient based on large amounts of calculations.

The earthquake effect coefficient α is defined as the ratio of the horizontal seismic action F to the gravity G of a single particle elastic system. The relationship among seismic effect coefficient, seismic action and gravity is:

$$F = \alpha G$$

The earthquake effect coefficient can be directly obtained from the Chinese *Code for Seismic Design of Buildings* (GB 50011—2010) Version 2016; thus, the horizontal seismic action of a single particle system can be easily derived.

Based on Eq.(3-23):

$$F = mS_a = mg \cdot \frac{S_a}{g} = \frac{S_a}{g} \cdot G$$

Thus, the earthquake effect coefficient can be represented as follows:

$$\alpha = \frac{S_a}{g} = \frac{|\ddot{x}_g|_{max}}{g} \cdot \frac{S_a}{|\ddot{x}_g|_{max}} \quad (3\text{-}24)$$

1. Earthquake Coefficient

In Eq.(3-24):

$$k = \frac{|\ddot{x}_g|_{max}}{g}$$

The parameter k is called the earthquake coefficient, which is the ratio between the peak ground motion acceleration and the acceleration due to gravity. The higher the seismic intensity, the bigger the ground motion acceleration, so is the seismic coefficient. Therefore, there is a corresponding relation between the seismic intensity and the seismic coefficient.

Currently, the relationship between the earthquake intensity and the earthquake coefficient used by Chinese seismic design and research groups is listed in Table 3-3.

Relationship between the earthquake intensity and the earthquake coefficient

Table 3-3

Earthquake intensity	VI(6)	VII(7)	VIII(8)	IX(9)
Earthquake coefficient k	0.05	0.1	0.2	0.4

2. Dynamic Coefficient

In Eq.(3-24):

$$\beta = \frac{S_a}{|\ddot{x}_g|_{max}} \quad (3\text{-}25)$$

The parameter β is called the dynamic coefficient, which is the ratio of the maximum absolute acceleration to the maximum ground motion acceleration for the SDOF system, reflecting how much the maximum absolute acceleration is amplified due to the dynamic effect.

3. Standard Earthquake Effect Coefficient Curve

When determining the earthquake effect coefficient α, the damping ratio is first set as 0.05. For a certain ground motion acceleration wave $\ddot{x}_g(t)$, α is only relevant to the structural natural vibration period T. In fact, for a given T, α can be calculated, which forms one-to-one relation and is called α–T curve. If another acceleration wave for ground motion is given, another α–T curve can be obtained. For a specified construction site, n suitable ground motion accelerations can be selected, and n curves can be obtained. A standard α–T curve for a specified construction site can be determined by analyzing and fitting these α–T curves in conjunction with appropriate adjustments based on engineering experience. This standard curve is also known as the standard earthquake effect coefficient curve (Fig. 3-9).

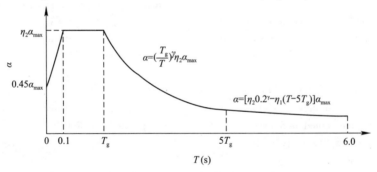

Fig. 3-9 Standard earthquake effect coefficient curve

When the damping ratio is not equal to 0.05, the curve can be obtained by correcting the standard earthquake effect coefficient curve (with a damping ratio of 0.05). The earthquake effect coefficient curves for seismic design provided by the Chinese *Code for Seismic Design of Buildings* (GB 50011—2010) Version 2016 are based on the above theory. Furthermore, details on using the earthquake effect coefficient curves to calculate the seismic action are introduced in Section 4.3.

3.5 Nonlinear Analysis

In the above analysis, it is assumed that a structure is in an elastic state and Rayleigh damping can be used; thus, the equations of motion are linear. In this section, material nonlinearity will be considered, although the Rayleigh damping assumption will still be used.

3.5.1 Nonlinearity of Materials

For reinforced concrete and masonry, which are two types of common materials in building structures, the stress-strain relationship is essentially linear for low-stress values. Macroscopically, the relationship between the forces or the restoring forces of the members and the deformations or displacements is linear. In the mechanical model, a member can be idealized as a linear spring. However, as the deformation increases, especially after the appearance of cracks, the stress-strain relationship is no longer linear because of unrecoverable deformation. Macroscopically, the relationship between the restoring force and the deformation of a member is not linear.

3.5.2 Equations of Motion of SDOF Nonlinear Systems $K(t)$

When the deformation of a system is relatively large, the stress-strain relationship of the material is no longer linear. Furthermore, the relationship between the restoring force and the deformation is no longer linear and can be given as follows:

$$f_s = f(x)$$

The form of the function $f(x)$ is relevant to the material properties and structural deformations.

Based on Eq. (3-4), the equation of motion for a nonlinear SDOF subjected to a ground motion acceleration can be given as follows:

$$m\ddot{x} + c\dot{x} + f(x) = -m\ddot{x}_g(t)$$

For convenience, the restoring force, which is the same as that for a linear system, can be expressed as follows:

$$f_s(x) = k(x) \cdot x$$

Where the coefficient $k(x)$ is associated with the deformation of the particle; in the linear system, it is a constant. Thus, the equation of motion for this system can be given as follows:

$$m\ddot{x} + c\dot{x} + k(x) \cdot x = -m\ddot{x}_g(t)$$

(3-26)

3.5.3 Solution of Nonlinear Motion Equation

For the equation of motion shown in Eq. (3-26), the numerical computational method presented in Section 3.4 is often used because the ground motion acceleration $\ddot{x}_g(t)$ is extremely variable.

For nonlinear systems, the equation of motion is valid at times t_{k-1} and t_k, and it can be given as follows:

$$m\ddot{x}(t_{k-1}) + c\dot{x}(t_{k-1}) + k[x(t_{k-1})]x(t_{k-1}) = -m\ddot{x}_g(t_{k-1})$$

$$m\ddot{x}(t_k) + c\dot{x}(t_k) + k[x(t_k)]x(t_k) = -m\ddot{x}_g(t_k)$$

By subtracting the former equation from the latter, assuming that:

$$\Delta x = x(t_k) - x(t_{k-1}),$$
$$\Delta \dot{x} = \dot{x}(t_k) - \dot{x}(t_{k-1}),$$
$$\Delta \ddot{x} = \ddot{x}(t_k) - \ddot{x}(t_{k-1}) \quad (3-27)$$
$$\Delta \ddot{x}_g = \ddot{x}_g(t_k) - \ddot{x}_g(t_{k-1}),$$
$$\Delta t = t_k - t_{k-1}$$

And that Δt is small enough to neglect the variation of the coefficient $k(x)$ in this time interval, we have:

$$m\Delta\ddot{x} + c\Delta\dot{x} + k\Delta x = -m\Delta\ddot{x}_g \quad (3-28)$$

Eq. (3-28) holds for any time interval $[t_{k-1}, t_k]$, $k=1,2,\cdots,n$, and the coefficient k is on average a constant value during the time interval.

Eq. (3-28) is called the equation of motion of a SDOF system in an incremental form. To solve Eq. (3-28), the linear acceleration method in an incremental form is used.

Based on Eqs. (3-18) and (3-19):

$$\Delta\dot{x} = \ddot{x}_{k-1} \cdot \Delta t + \frac{1}{2}\Delta\ddot{x} \cdot \Delta t \quad (3-29)$$

$$\Delta x = \dot{x}_{k-1} \cdot \Delta t + \frac{1}{2}\ddot{x}_{k-1} \cdot \Delta t^2 + \frac{1}{6}\Delta\ddot{x} \cdot \Delta t^2$$

(3-30)

Based on Eq.(3-30):

$$\Delta \ddot{x} = \frac{6}{\Delta t^2}\Delta x - \frac{3}{\Delta t}(2\dot{x}_{k-1} + \ddot{x}_{k-1} \cdot \Delta t)$$

(3-31)

By substituting Eq.(3-31) into Eq.(3-30), we can obtain:

$$\Delta \dot{x} = \ddot{x}_{k-1} \cdot \Delta t + \frac{3\Delta x}{\Delta t} - 3(\dot{x}_{k-1} + \frac{1}{2}\ddot{x}_{k-1} \cdot \Delta t)$$

$$= \frac{3}{\Delta t}\Delta x - 3\dot{x}_{k-1} - \frac{1}{2}\ddot{x}_{k-1} \cdot \Delta t \quad (3\text{-}32)$$

By substituting Eq.(3-31) and Eq.(3-32) into Eq.(3-28), we can obtain:

$$m\frac{6}{\Delta t^2}\Delta x - \frac{3m}{\Delta t}(2\dot{x}_{k-1} + \ddot{x}_{k-1}\Delta t)$$
$$+ \frac{3c}{\Delta t}\Delta x - c(3\dot{x}_{k-1} + \frac{1}{2}\ddot{x}_{k-1} \cdot \Delta t)$$
$$+ k\Delta x = - m\Delta \ddot{x}_g$$

Lastly, the equation can be arranged as follows:

$$\left(m\frac{6}{\Delta t^2} + c \cdot \frac{3}{\Delta t} + k\right)\Delta x = -m\Delta \ddot{x}_g +$$
$$\frac{3m}{\Delta t}(2\dot{x}_{k-1} + \ddot{x}_{k-1}\Delta t) + c\left(3\dot{x}_{k-1} + \frac{1}{2}\ddot{x}_{k-1} \cdot \Delta t\right)$$

(3-33)

Let $k^* = m\frac{6}{\Delta t^2} + c \cdot \frac{3}{\Delta t} + k$

$$\Delta P^* = - m\Delta \ddot{x}_g + \frac{3m}{\Delta t}(2\dot{x}_{k-1} + \ddot{x}_{k-1}\Delta t)$$
$$+ c(3\dot{x}_{k-1} + \frac{1}{2}\ddot{x}_{k-1} \cdot \Delta t)$$

Then, Eq.(3-33) can be written as follows:

$$k^* \Delta x = \Delta P^* \quad (3\text{-}34)$$

where,

k^* ——the quasi-stiffness of the system;

ΔP^* ——the quasi-load increment.

The calculation process can be described in the following manner. First, determine Δx from Eq.(3-34). Second, determine $\Delta \dot{x}$ and $\Delta \ddot{x}$ by substituting Δx into Eq.(3-29) and Eq.(3-31). Then, determine the displacement, velocity and acceleration of the system at time t_k. Lastly, determine the displacement, velocity and acceleration of the system at times t_{k+1}, t_{k+2}, $\cdots$, t_n by repeating the above steps.

3.5.4 Hysterestic Model

When solving the equation of motion for a nonlinear system, it can be assumed that the stiffness coefficient $k(x)$ changes minimally during the time interval Δt. Hence, $k(x)$ is considered a constant. However, $k(x)$ varies during different time intervals. Therefore, it is necessary to study the restoring force model of the system to determine $k(x)$.

The restoring force model reflects the changing trends in the restoring force f_s during the whole vibration process. The restoring forces vary in a complex manner. Fig. 3-10 illustrates the relationship between the restoring force and the deformation of a piece of masonry wall measured under a lateral force. To study the nonlinear seismic response of the building structures, it is necessary to establish a practical restoring force model based on a large amount of data.

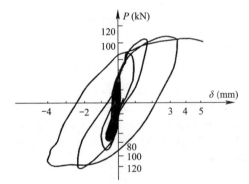

Fig. 3-10 Relationship between the restoring force and deformation

At present, scholars at home and abroad have proposed some practical restoring force models for reinforced concrete and masonry members. In this section, a semi-deteriorated trilinear restoring force model for the nonlinear seismic responses of masonry structures will be introduced.

The semi-deteriorated trilinear restoring force model is shown in Fig. 3-11. The parameter P is the calculated seismic shear force, P_x is the cracking strength, and P_y is the yield strength.

When $P \leqslant P_x$, the loading and unloading stiffness is assumed to be K_1. When $P_x \leqslant P \leqslant P_y$, the loading stiffness is K_2, the unloading stiffness is taken as K_1, and the calculated seismic shear force should be corrected. When $P \geqslant P_y$, the loading stiffness is taken as K_3, the unloading stiffness is taken as K_{12}, and the calculated seismic shear force should be corrected. When the deformation is in the opposite direction, the loading and unloading stiffness conform to a similar principle.

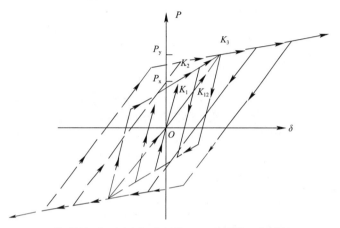

Fig. 3-11 Semi-deteriorated trilinear restoring force model

3.6 Free Vibration of Multi-Degree-of-Freedom Systems

In general, the mass of an actual structure is continuously distributed; therefore, its dynamic degrees of freedom are infinite. However, a practical dynamic analysis of a model with infinite degrees of freedom is too complicated to perform; furthermore, the use of a model with a finite degree of freedom is accurate for general engineering designs.

A structure can be represented as a SDOF system and its dynamic response can be evaluated by solving the equation of motion. Moreover, seismic response analysis and seismic design of SDOF systems are extremely significant in practice. Nevertheless, there are a number of structural types in practical engineering, such as multi-storey and high-rise buildings, which should not be mathematically reduced to SDOF models, i. e. , a SDOF model cannot adequately describe the response of a structure. In addition, the failure pattern of certain structures can only be described by more than one degree of freedom. Therefore, for multi-storey and high-rise buildings, a multi-degree-of-freedom model is generally used. For practical structures, the mass of each floor or roof can be combined, and the mass of the walls and columns between the floors or between a floor and the roof is combined and simplified to be a single mass for the floor or roof. This type of multi-degree-of-freedom model is called an inter-storey model in engineering; a diagram is depicted in Fig. 3-12.

3.6.1 Equations of Motion

As an example, we use a two-degree-of-freedom storey shear model, for which the computational model is depicted in Fig. 3-13. We can directly formulate the equation of motion of the system by expressing the equilibrium of all forces acting on the mass according to D'Alembert's principle.

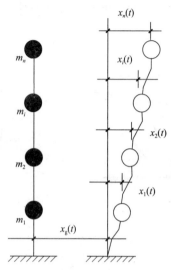

Fig. 3-12 Diagram of an inter-storey model

Based on the diagram illustrated in Fig. 3-13(c), the inertia force f_{I1} and the restoring force f_{S1} of mass 1 can be given as follows:
$$f_{I1} = -m_1[\ddot{x}_1(t) + \ddot{x}_g(t)]$$
$$f_{S1} = -f_{S11} + f_{S12} = -K_1 x_1(t) + K_2[x_2(t) - x_1(t)]$$

According to D'Alembert's principle, the equilibrium of the forces is:
$$f_{I1} + f_{S1} = -m_1\ddot{x}_1(t) - m_1\ddot{x}_g(t) - K_1 x_1(t) + K_2 x_2(t) - K_2 x_1(t) = 0$$

Furthermore, the equation can be rewritten as follows:
$$m_1\ddot{x}_1(t) + (K_1 + K_2)x_1(t) - K_2 x_2(t) = -m_1\ddot{x}_g(t) \tag{3-35}$$

Based on the diagram illustrated in Fig. 3-13(c), the inertia force f_{I2} and restoring force f_{S2} of mass 2 can be represented as follows:
$$f_{I2} = -m_2[\ddot{x}_2(t) + \ddot{x}_g(t)]$$
$$f_{S2} = -f_{S21} = -K_2[x_2(t) - x_1(t)]$$

where,

$\ddot{x}_g(t)$——the ground motion acceleration;

$x_1(t), x_2(t)$——the displa-cements of m_1 and m_2 relative to the base, respectively;

$\ddot{x}_1(t), \ddot{x}_2(t)$——the accelerations of m_1 and m_2, respectively;

$\ddot{x}_1(t)+\ddot{x}_g(t), \ddot{x}_2(t)+\ddot{x}_g(t)$——the absolute accelerations of m_1 and m_2, respectively.

According to D'Alembert's principle, the equilibrium of the forces is:
$$f_{I2} + f_{S2} = -m_2\ddot{x}_2(t) - m_2\ddot{x}_g(t) - K_2[x_2(t) - x_1(t)] = 0$$

This equation can be rewritten as follows:
$$m_2\ddot{x}_2(t) + K_2 x_2(t) - K_2 x_1(t) = -m_2\ddot{x}_g(t) \tag{3-36}$$

Combine the two equations, and describe it by the matrix:
$$\begin{bmatrix} m_1 & 0 \\ 0 & m_2 \end{bmatrix} \begin{Bmatrix} \ddot{x}_1(t) \\ \ddot{x}_2(t) \end{Bmatrix} + \begin{bmatrix} K_1 + K_2 & -K_2 \\ -K_2 & K_2 \end{bmatrix} \begin{Bmatrix} x_1(t) \\ x_2(t) \end{Bmatrix} = -\begin{bmatrix} m_1 & 0 \\ 0 & m_2 \end{bmatrix} \begin{Bmatrix} \ddot{x}_g(t) \\ \ddot{x}_g(t) \end{Bmatrix}$$

Assume:
$$[M] = \begin{bmatrix} m_1 & 0 \\ 0 & m_2 \end{bmatrix}$$
$$[K] = \begin{bmatrix} K_1 + K_2 & -K_2 \\ -K_2 & K_2 \end{bmatrix}$$
$$\{x(t)\} = \begin{Bmatrix} x_1(t) \\ x_2(t) \end{Bmatrix}$$
$$\{\ddot{x}(t)\} = \begin{Bmatrix} \ddot{x}_1(t) \\ \ddot{x}_2(t) \end{Bmatrix}$$
$$\{I\} = \begin{Bmatrix} 1 \\ 1 \end{Bmatrix}$$

The equation of motion for a 2-DOF system can be written in matrix form as follows:
$$[M]\{\ddot{x}(t)\} + [K]\{x(t)\} = -[M]\{I\}\ddot{x}_g(t) \tag{3-37}$$

where,

$[M]$——the mass matrix;

$[K]$——the stiffness matrix;

$\{\ddot{x}(t)\}, \{x(t)\}$—— the acceleration vector and the displacement vector, respectively.

If we consider the impact of damping, then the equation of motion is as follows:
$$[M]\{\ddot{x}(t)\} + [C]\{\dot{x}(t)\} + [K]\{x(t)\} = -[M]\{I\}\ddot{x}_g(t) \tag{3-38}$$

where,

$[C]$——the damping matrix, and its specific form is related to the assumption of damping. Generally, the Rayleigh damping assumption is applied. In the Rayleigh damping assumption, the damping matrix $[C]$ is defined as follows:

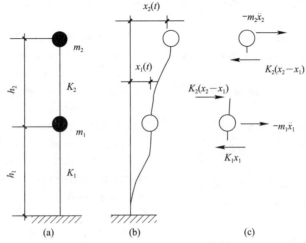

Fig. 3-13 Diagram of a storey shear model with two degrees of freedom
(a) Calculation model; (b) Displacement of two particles; (c) Force balance of two particles

$$[C] = \alpha_0 [M] + \alpha_1 [K]$$

where,

α_0, α_1——the constants related to the structural system.

For general MDOF (multi degrees of freedom) systems, the equation of motion is similar to the form of Eq. (3-38); however, the mass matrix, the damping matrix and the stiffness matrix vary according to structural forms and mechanical models.

For multi-storey and high-rise buildings, an inter-storey model is preferred, which assumes that only rigid-body motion occurs when the floors vibrate. Inter-storey models can be classified by the storey deformation characteristics as follows: (1) storey shear model, which only considers shear deformation; (2) storey bending model, which only considers bending deformation; and (3) storey bending shear model, which considers both bending and shear deformation simultaneously.

3.6.2 Vibration Properties of Multi-Degree-of-Freedom Systems

1. Frequency

If we leave the force applied in Eq. (3-38) equal to zero and neglect the influence of damping, the equation of motion for an undamped MDOF system is as follows:

$$[M]\{\ddot{x}(t)\} + [K]\{x(t)\} = 0$$
(3-39)

For example, take a 2-DOF system. Then the displacement vector can be defined as follows:

$$\{x(t)\} = \begin{Bmatrix} x_1(t) \\ x_2(t) \end{Bmatrix} = \begin{Bmatrix} X_1 \\ X_2 \end{Bmatrix} \sin(\omega t + \psi)$$
(3-40)

where,

X_1, X_2——the amplitudes of the displacement of m_1 and m_2, respectively.

By introducing Eq. (3-40) into Eq. (3-39), we can obtain the following:

$$- [M] \begin{Bmatrix} X_1 \\ X_2 \end{Bmatrix} \omega^2 \sin(\omega t + \varphi)$$

$$+ [K] \begin{Bmatrix} X_1 \\ X_2 \end{Bmatrix} \sin(\omega t + \varphi) = 0$$

Then, this equation can be converted to:

$$([K] - \omega^2 [M]) \begin{Bmatrix} X_1 \\ X_2 \end{Bmatrix} = 0 \quad (3-41)$$

Eq. (3-41) is a homogeneous system of linear algebraic equations. It is clear that $X_1 = 0$, $X_2 = 0$ is a group of solutions. However, Eq. (3-41) indicates that the displacement terms $x_1(t)$ and

$x_2(t)$ could both be zero if $X_1 = X_2 = 0$, which implies no vibration occurs at all. Therefore, it is not the solution for the free vibration condition.

To obtain a nonzero solution to Eq.(3-41), the determinant of the coefficient matrix should be zero, which leads to:

$$|[K] - \omega^2[M]| = 0 \quad (3-42)$$

or

$$\begin{vmatrix} K_1 + K_2 - m_1\omega^2 & -K_2 \\ -K_2 & K_2 - m_2\omega^2 \end{vmatrix} = 0 \quad (3-43)$$

The above equations are called frequency equations, then we can obtain the solution of ω^2 as follows:

$$(\omega^2)^2 - \left(\frac{K_1 + K_2}{m_1} + \frac{K_2}{m_2}\right)\omega^2 + \frac{K_1 K_2}{m_1 m_2} = 0$$

By solving these equations, we can obtain the following:

$$\omega^2 = \frac{1}{2}\left(\frac{K_1 + K_2}{m_1} + \frac{K_2}{m_2}\right) \pm \sqrt{\frac{1}{4}\left(\frac{K_1 + K_2}{m_1} + \frac{K_2}{m_2}\right)^2 - \frac{K_1 K_2}{m_1 m_2}} \quad (3-44)$$

Alternatively, this equation can be written in the following form:

$$\omega^2 = \frac{1}{2}\left[\frac{K_1 + K_2}{m_1} + \frac{K_2}{m_2} \pm \sqrt{\left(\frac{K_1}{m_1} - \frac{K_2}{m_2}\right)^2 + \frac{K_2}{m_1}\left(\frac{2K_1 + K_2}{m_1} + \frac{2K_2}{m_2}\right)}\right] \quad (3-45)$$

When the stiffness, mass and quantity under the radical in Eq.(3-45) are positive and the first term in Eq.(3-44) exceeds the second, we can obtain the two real solutions with positive-sign, which are the natural circular frequencies of the 2-DOF systems. Here, ω_1, which is the smaller one, is called the first natural circular frequency or the basic natural circular frequency, whereas ω_2, which is the larger one, is called the second natural circular frequency. Similar to the case of SDOF systems, $f_1 = \omega_1/2\pi$ is called the first natural frequency or the basic natural frequency, $T_1 = 2\pi/\omega_1$ is called the first natural period or basic natural period, $f_2 = \omega_2/2\pi$ is called the second modal frequency and $T_2 = 2\pi/\omega_2$ is called the second natural period.

2. Vibration Mode and Mode Shape

By introducing ω_1 and ω_2 into Eq.(3-41), we can obtain the displacement amplitudes of m_1 and m_2. After introducing ω_1 or ω_2 into Eq.(3-41), the determinant of the coefficient is zero; therefore, the solutions for X_1 and X_2 are not unique, and the relationship between X_1 and X_2 can be obtained. For ω_1, we can define X_{11} and X_{12} as the solutions for X_1 and X_2, and for ω_2, we can define X_{21} and X_{22} as the solutions for X_1 and X_2. Then, by introducing ω_1 and ω_2 into Eq.(3-41), we can obtain the following:

For ω_1: $\dfrac{X_{12}}{X_{11}} = \dfrac{K_1 + K_2 - m_1\omega_1^2}{K_2}$

For ω_2: $\dfrac{X_{22}}{X_{21}} = \dfrac{K_1 + K_2 - m_1\omega_2^2}{K_2}$

By solving Eq.(3-40), the system displacements under free vibration conditions can be obtained as follows:

For ω_1: $\begin{Bmatrix} x_{11}(t) \\ x_{12}(t) \end{Bmatrix} = \begin{Bmatrix} X_{11} \\ X_{12} \end{Bmatrix} \sin(\omega_1 t + \psi_1)$

For ω_2: $\begin{Bmatrix} x_{21}(t) \\ x_{22}(t) \end{Bmatrix} = \begin{Bmatrix} X_{21} \\ X_{22} \end{Bmatrix} \sin(\omega_2 t + \psi_2)$

When the system vibrates at the first or second natural frequencies, the displacement ratios of the two particles are as follows:

For ω_1: $\dfrac{x_{12}(t)}{x_{11}(t)} = \dfrac{X_{12}}{X_{11}} = \dfrac{K_1 + K_2 - m_1\omega_1^2}{K_2}$

For ω_2: $\dfrac{x_{22}(t)}{x_{21}(t)} = \dfrac{X_{22}}{X_{21}} = \dfrac{K_1 + K_2 - m_1\omega_2^2}{K_2}$

Thus, the ratios are constant and are not related to time t, i.e., the displacement ratio of the two particles remains invariant when the system vibrates with its natural frequencies. This particular vibration form is called a vibration mode. When the system vibrates at ω_1, it is called the first mode or the fundamental mode, and when the system vibrates at ω_2, it is called the second mode.

In general, a n-DOF system will have n frequencies and n modes, these particular vibration forms are called vibration modes. It is an inherent characteristic of the system. Because the displacement ratios and the velocity ratios of these particles, within one vibration mode, are constants, this type of vibration pattern occurs only when the initial ratios of the displacements and the velocities of the particles are the same as that of the main vibration modes, i. e., for a given initial condition.

Under a general initial condition, the vibration curve of the system would contain all of the vibration mode shapes. Considering the characteristics of homogeneous linear equations, its general solution is a linear combination of all linear independent particular solutions. In terms of 2-DOF systems, the general solution can be written as follows:

$$\{x(t)\} = \begin{Bmatrix} x_1(t) \\ x_2(t) \end{Bmatrix} = \begin{Bmatrix} X_{11} \\ X_{12} \end{Bmatrix} \sin(\omega_1 t + \psi_1)$$
$$+ \begin{Bmatrix} X_{21} \\ X_{22} \end{Bmatrix} \sin(\omega_2 t + \psi_2)$$

From the equation above, under general initial conditions, the vibration of any particles is a composite of the simple harmonic vibrations for all of the main vibration modes. It does not remain a simple harmonic vibration, and the displacement ratios of all particles vary with time instead of being constant.

3. Orthogonality of the Vibration Modes

The so-called orthogonality of modes refers to the mutually orthogonal property between two main vibration modes of different frequencies in MDOF systems.

We can define ω_i as the i-th frequency, and the corresponding vibration mode is $\{X\}_i$; if ω_j is the j-th frequency, then, the corresponding vibration mode is $\{X\}_j$. Based on the characteristics of the frequencies and vibration modes, any natural frequencies and vibration modes should satisfy Eq. (3-41), specifically:

$$([K] - \omega_i^2[M])\{X\}_i = 0 \quad (3-46)$$
$$([K] - \omega_j^2[M])\{X\}_j = 0 \quad (3-47)$$

By pre-multiplying Eqs. (3-46) and (3-47) with $\{X\}_j^T$ and $\{X\}_i^T$, we obtain the following:

$$\{X\}_j^T([K] - \omega_i^2[M])\{X\}_i = 0 \quad (3-48)$$

$$\{X\}_i^T([K] - \omega_j^2[M])\{X\}_j = 0 \quad (3-49)$$

We can transpose the matrix on the left side of Eq. (3-48), and the transpose of the matrix on the right side equals zero; thus, we have:

$$\{X\}_i^T([K]^T - \omega_i^2[M]^T)\{X\}_j = 0 \quad (3-50)$$

For general building construction, the stiffness and mass matrices are symmetric. Thus, we have:

$$[K]^T = [K], [M]^T = [M]$$

Eq. (3-50) can be rewritten as follows:

$$\{X\}_i^T([K] - \omega_i^2[M])\{X\}_j = 0 \quad (3-51)$$

By subtracting Eq. (3-49) from Eq. (3-51), we can obtain the following:

$$(\omega_j^2 - \omega_i^2)\{X\}_i^T[M]\{X\}_j = 0$$

By considering that $\omega_i \neq \omega_j$, we arrive at:

$$\{X\}_i^T[M]\{X\}_j = 0 \quad (3-52)$$

Eq. (3-52) is called the first orthogonal condition of the vibration modes, i. e., the orthogonal condition or weighted orthogonal condition of the vibration modes for the mass matrix.

By introducing Eq. (3-52) into Eq. (3-51), we get:

$$\{X\}_i^T[K]\{X\}_j = 0 \quad (3-53)$$

Eq. (3-53) is called the second orthogonal condition of the vibration modes, which is the orthogonal condition or weighted orthogonal condition of the vibration modes for the stiffness matrix.

3.6.3 Approximation Method for the Natural Properties

As presented in Section 3.6.2, the natural frequencies and the vibration modes can be obtained by solving the frequency equation. However, the manual calculation is excessively tedious and rather complex because the system has too many degrees of freedom. Therefore, either computer programs or approximations are generally used to address such practical engineering calculations. In this section, a generalized Jacobi method is presented, which is commonly used to determine structures' natural frequencies and vibration modes. Some approximation methods are also presented, such as the matrix iteration method (Stodola method), the energy method (Rayleigh method), the equivalent mass method (Dunkerley formula), etc. Please refer to Ref. [8] for additional details.

1. Generalized Jacobi Method

For MDOF systems, the problem of solving for the vibration frequencies and vibration modes can be narrowed down to the solution of the eigenvalues λ_i and eigenvectors $\{X\}$ of the following equation:

$$[K]\{X\} = \lambda[M]\{X\} \quad (3\text{-}54)$$

The basic idea of the generalized Jacobi method is to determine appropriate similarity transformation matrices $[P]$ and $[Q]$ by letting:

$$[P][K][Q] = \begin{bmatrix} \alpha_1 & & \\ & \alpha_i & \\ & & \alpha_n \end{bmatrix},$$

$$[P][M][Q] = \begin{bmatrix} \beta_1 & & \\ & \beta_i & \\ & & \beta_n \end{bmatrix}$$

Then, the eigenvalues are $\lambda_i = \alpha_i/\beta_i (i = 1, 2, 3, \cdots, n)$.

Because of the similarity transformation, the eigenvalues of Eq. (3-54) are equal to the eigenvalues obtained above.

2. Matrix Iteration Method (Stodola Method)

The matrix iteration method assumes the mode shapes and then attempts to perform iteration and adjustment steps until satisfactory results are obtained. Finally, the natural frequency is determined.

Let us assume that the inverse matrix of the system stiffness matrix is $[K]^{-1}$. By pre-multiplying Eq. (3-54) with $[K]^{-1}$, we obtain the following:

$$\frac{1}{\lambda}\{X\} = [K]^{-1}[M]\{X\} \quad (3\text{-}55)$$

Given $[D] = [K]^{-1}[M]$, then:

$$[D]\{X\} = \frac{1}{\lambda}\{X\} \quad (3\text{-}56)$$

Eq. (3-56) is the iterative equation that must be solved, and the matrix $[D]$ represents all of the dynamic characteristics of a structure; therefore, $[D]$ is called the dynamic matrix.

The iteration steps can be defined as follows:

(1) First, assume a tentative vibration-mode $\{\overline{X}_1^0\}$, where subscript 1 indicates the first mode, and superscript 0 represents the initial vibration mode. By introducing $\{\overline{X}_1^0\}$ into Eq. (3-56), we have:

$$[D]\{\overline{X}_1^0\} = \frac{1}{\lambda}\{\overline{X}_1^1\} = \{\hat{X}_1^1\} \quad (3\text{-}57)$$

(2) Generally, $\{\overline{X}_1^0\}$ and $\{\overline{X}_1^1\}$ are different unless $\{\overline{X}_1^0\}$ is the actual vibration mode, i.e., $\{\overline{X}_1^0\}$ and $\{\overline{X}_1^1\}$ are disproportionate. By introducing $\{\overline{X}_1^1\}$ into Eq. (3-56), we can obtain the following:

$$[D]\{\overline{X}_1^1\} = \frac{1}{\lambda}\{\overline{X}_1^2\} = \{\hat{X}_1^2\} \quad (3\text{-}58)$$

where,

$\{\overline{X}_1^2\}$ ——the second approximate vibration modal vector.

(3) If the error between $\{\overline{X}_1^2\}$ and $\{\overline{X}_1^1\}$ satisfies the requirement, then λ in Eq. (3-58) is the desired eigenvalues; otherwise, the iteration needs to be continued. It can be proven that the iterative process will eventually converge to the first mode; the specific theory is available in Ref. [8].

3. Energy Method (Rayleigh Method)

We can assume that $[K]$ and $[M]$ in Eq. (3-54) are both positive-definite matrices; thus, all of the eigenvalues λ are positive.

We can assume that λ_r and $\{\overline{X}\}_r$ are the r-order eigenvalues and their corresponding eigenvectors, respectively:

$$[K]\{\overline{X}\}_r = \lambda_r[M]\{\overline{X}\}_r \quad (3\text{-}59)$$

By pre-multiplying Eq. (3-59) with $\frac{1}{2}\{\overline{X}\}_r^T$, we get:

$$\frac{1}{2}\lambda_r\{\overline{X}\}_r^T[M]\{\overline{X}\}_r = \frac{1}{2}\{\overline{X}\}_r^T[K]\{\overline{X}\}_r \quad (3\text{-}60)$$

It should be noted that $\lambda_r = \omega_r^2$. In fact, Eq. (3-60) shows that the maximum kinetic energy equals the maximum potential energy when the systems are set into free vibration with a r-order vibration mode. The following can then be obtained from Eq. (3-60):

$$\omega_r^2 = \lambda_r = \frac{\{\overline{X}\}_r^T[K]\{\overline{X}\}_r}{\{\overline{X}\}_r^T[M]\{\overline{X}\}_r} \quad (3\text{-}61)$$

Now, if we let $\{X\}$ be an arbitrary displacement column vector, then:

$$R(x) = \frac{\{X\}^T[K]\{X\}}{\{X\}^T[M]\{X\}} \quad (3\text{-}62)$$

where $R(x)$ is referred to as the Rayleigh quotient. From Eq. (3-62), we know that $R(x)$ is a number that is related not only to the mass matrix and the stiffness matrix of the system, but also to the hypothetical displacement vector. If the selected displacement vector is equal to a certain characteristic vector of the system, then $R(x)$ is the eigenvalue corresponding to the characteristic vector.

The basic method of computing the eigenvalue of the system using the Rayleigh quotient is selecting a displacement vector to approximate the eigenvectors based on the boundary conditions. Furthermore, the calculated Rayleigh quotient is the approximated eigenvalue of this eigenvector. Generally, the Rayleigh quotient is used to calculate the first vibration frequency, and it can be proven that the Rayleigh quotient is always greater than the first eigenvalue of the system.

4. Equivalent Mass Method (Dunkerley Formula)

The fundamental frequency of a structure determined by the Rayleigh method is always greater than or equal to the actual fundamental frequency, i. e., it is the upper limit of the structure's fundamental frequency. However, the Dunkerley formula introduced in this section can be used to evaluate the lower limit of the structure's fundamental frequency.

The Dunkerley formula is based on the relationship between the roots and the coefficients of the following algebraic equation.

Given an equation of the n-th order:

$$x^n + a_1 x^{n-1} + a_2 x^{n-2} + \cdots + a_i x^{n-i} + \cdots + a_{n-1} x + a_n = 0 \quad (3\text{-}63)$$

The roots are $x_1, x_2, x_3, \cdots, x_n$, and Eq. (3-63) can be written as follows:

$$(x - x_1)(x - x_2)(x - x_3)\cdots(x - x_n) = 0 \quad (3\text{-}64)$$

Based on the relationship between the roots and the coefficients, we have:

$$\sum_{i=1}^{n} x_i = -a_1 \quad (3\text{-}65)$$

$$\sum_{i=1}^{n}\sum_{\substack{j=1\\j\neq i}}^{n} x_i x_j = 2a_2 \quad (3\text{-}66)$$

Moreover, we have the following:

$$\sum_{i=1}^{n} x_i^2 = \left(\sum_{i=1}^{n} x_i\right)^2 - \sum_{i=1}^{n}\sum_{\substack{j=1\\j\neq i}}^{n} x_i x_j = a_1^2 - 2a_2 \quad (3\text{-}67)$$

To ensure that $\{X\}$ has a nonzero solution, the determinant of the coefficient must be zero:

$$\begin{bmatrix} \frac{1}{\lambda} - M_1 f_{11} & -M_2 f_{12} & \cdots & -M_n f_{1n} \\ -M_1 f_{21} & \frac{1}{\lambda} - M_2 f_{22} & \cdots & -M_n f_{2n} \\ \vdots & \vdots & \vdots & \vdots \\ -M_1 f_{n1} & -M_2 f_{n2} & \cdots & \frac{1}{\lambda} - M_n f_{nn} \end{bmatrix} = 0$$

where,

$\lambda = \omega^2$.

Eq. (3-54) can be rewritten as follows:

$$\left(\frac{1}{\lambda}[I] - [K]^{-1}[M]\right)\{X\} = 0 \quad (3\text{-}68)$$

where,

$$[K]^{-1} = [F] = \begin{bmatrix} f_{11} & f_{12} & \cdots & f_{1n} \\ f_{21} & f_{22} & \cdots & f_{2n} \\ \vdots & \vdots & \vdots & \vdots \\ f_{n1} & f_{n2} & \cdots & f_{nn} \end{bmatrix};$$

$[I]$——a unit matrix.

By expanding the above determinant, we get:
$$\frac{1}{\lambda^n} - \sum_{i=1}^{n} M_i f_{ii} \frac{1}{\lambda^{n-1}} +$$
$$\left[\frac{1}{2}\sum_{i=1}^{n}\sum_{j=1}^{n}(M_i f_{ii} M_j f_{jj} - M_i M_j f_{ij}^2)\right]$$
$$\frac{1}{\lambda^{n-2}} + \cdots = 0$$

By combining Eqs. (3-65) and (3-67), we have the following:
$$\sum_{i=1}^{n}\frac{1}{\omega_i^2} = \sum_{i=1}^{n} M_i f_{ii} \quad (3\text{-}69)$$

and
$$\sum_{i=1}^{n}\frac{1}{\omega_i^4} = \left(\sum_{i=1}^{n} M_i f_{ii}\right)^2 -$$
$$\sum_{i=1}^{n}\sum_{\substack{j=1\\j\neq i}}^{n}(M_i M_j f_{ii} f_{jj} - M_i M_j f_{ij}^2) \quad (3\text{-}70)$$

Given:
$$\frac{1}{f_{ii}} = K_{ii}$$

Then:
$$M_i f_{ii} = \frac{M_i}{K_{ii}} = \frac{1}{\omega_i^2} = \lambda_i \quad (3\text{-}71)$$

Thus, Eq. (3-69) can be written as follows:
$$\frac{1}{\omega_1^2} + \frac{1}{\omega_2^2} + \frac{1}{\omega_3^2} + \cdots + \frac{1}{\omega_n^2} =$$
$$\frac{1}{\omega_{11}^2} + \frac{1}{\omega_{22}^2} + \frac{1}{\omega_{33}^2} + \cdots + \frac{1}{\omega_{nn}^2} \quad (3\text{-}72)$$

Assume that $\omega_1 < \omega_2 < \omega_3 < \cdots < \omega_n$. In the calculation of ω_1, if the impact of the high-frequency term is neglected, the first frequency can be approximated as follows:
$$\frac{1}{\omega_1^2} \approx \sum_{i=1}^{n}\frac{1}{\omega_{ii}^2} \quad (3\text{-}73)$$

Eq. (3-73) is the desired Dunkerley formula for the first frequency. Obviously, the first frequency evaluated by the Dunkerley formula is smaller than the actual one.

5. Rayleigh-Ritz Method

Ritz promoted the Rayleigh method and found a better approach to obtain the first several mode shapes, i.e., the Rayleigh-Ritz method.

First, a group of linearly independent Ritz vectors that satisfy the initial constraints is selected. Generally, selecting $2s$ vectors when the first frequencies and vibration modes require a high accuracy is appropriate.

The displacement vector can be represented by a set of assumed shape vectors and a set of generalized coordinates of unknown magnitude, i.e., $X_i(i=1,2,3,\cdots,2s)$ and $q_i(i=1,2,3,\cdots,2s)$, where:
$$\{x\} = [X]\{q\}$$
or
$$\{x\} = \{X_1\}q_1 + \{X_2\}q_2 + \{X_3\}q_3 + \cdots \quad (3\text{-}74)$$

The maximum kinetic energy and potential energy of the system are defined as follows:
$$T_{\max} = \frac{1}{2}\omega^2\{q\}^T[X]^T[M][X]\{q\}$$
$$V_{\max} = \frac{1}{2}\{q\}^T[X]^T[K][X]\{q\}$$

Given $T_{\max} = V_{\max}$, the Rayleigh quotient can be obtained as follows:
$$R(x) = \hat{\omega}^2 = \frac{\{q\}^T[\hat{K}]\{q\}}{\{q\}^T[\hat{M}]\{q\}} \quad (3\text{-}75)$$

where,
$$[\hat{K}] = [X]^T[K][X] \text{——the generalized stiffness;}$$
$$[\hat{M}] = [X]^T[M][X] \text{——the generalized mass.}$$

Both $[\hat{K}]$ and $[\hat{M}]$ are $2s \times 2s$ square matrices.

When the Ritz vectors are selected, the optimal value of q should make $\hat{\omega}^2$ reach a minimum value, which can be expressed as follows:
$$\frac{\partial R(x)}{\partial q_i} = 0 \quad (i = 1,2,\cdots,2s)$$
$$(3\text{-}76)$$

Additionally, the following is also valid:
$$[\hat{K}]\{q\} = R(x)[\hat{M}]\{q\} = \hat{\omega}^2[\hat{M}]\{q\}$$
$$(3\text{-}77)$$

Thus, the n-DOF system is transformed into a $2s$-degree-of-freedom system $(2s<n)$. By solving Eq. (3-77), we can obtain $2s$ eigenvalues, $\hat{\omega}^2$, and the corresponding eigenvectors, $\{X\}_i (i=1,2,3,\cdots,2s)$. Based on Eq. (3-74), the approximate solution for the first $2s$ eigenvectors (vibration modes) can be obtained.

3.7 Modal Analysis Method for Multi-Degree-of-Freedom Systems

As mentioned previously, the equation of motion for MDOF systems is a group of coupled differential equations under the influence of ground motion, which leads to computational challenges. By using the orthogonality of modes, the differential equation set of a MDOF system can be decoupled into independent differential equations of n SDOF systems. Consequently, the original dynamic computing problem for MDOF systems is broken into several SDOF systems. After solving the responses of n SDOF systems, the seismic response of the MDOF systems can be obtained using a combination of the solutions of the SDOF systems, such as (1) sum of the absolute values; (2) SRSS (square root of the sum of the squares); (3) CQC (complete quadratic combination).

For simplicity, we will use a 2-DOF system as an example. The displacements of particles m_1 and m_2 under seismic action at any one time t, i.e., $x_1(t)$ and $x_2(t)$, can be expressed by the linear combinations of their two vibration modes:

$$\begin{aligned} x_1(t) &= q_1(t)X_{11} + q_2(t)X_{21} \\ x_2(t) &= q_1(t)X_{12} + q_2(t)X_{22} \end{aligned} \quad (3\text{-}78)$$

Eq. (3-78) is an equation for the coordinate transformation. The original variables $x_1(t)$ and $x_2(t)$ are in geometric coordinates, and the new coordinates $q_1(t)$ and $q_2(t)$ are generalized coordinates. Because the vibration mode of the system can be uniquely determined, the displacements of particles $x_1(t)$ and $x_2(t)$ can be subsequently determined when $q_1(t)$ and $q_2(t)$ are determined.

Eq. (3-78) indicates that the displacement of the system can be regarded as the superposition of the vibration modes multiplied by the corresponding coefficients $q_1(t)$ and $q_2(t)$. In fact, this method decomposes the actual displacement by the vibration modes, which is known as the modal superposition method.

For MDOF systems, Eq. (3-78) can be written in matrix form as follows:

$$\{x(t)\} = [X]\{q(t)\} \quad (3\text{-}79)$$

where,

$$\{x(t)\} = \begin{Bmatrix} x_1(t) \\ x_2(t) \\ \vdots \\ x_i(t) \\ \vdots \\ x_n(t) \end{Bmatrix}; \quad \{q(t)\} = \begin{Bmatrix} q_1(t) \\ q_2(t) \\ \vdots \\ q_i(t) \\ \vdots \\ q_n(t) \end{Bmatrix};$$

$$[X] = \{\{X\}_1 \quad \{X\}_2 \quad \cdots \quad \{X\}_n\};$$

$\{x(t)\}$ ——the displacement vector;

$\{q(t)\}$ ——the generalized coordinate vector;

$[X]$ ——the vibration mode matrix, in which $\{X\}_i$, i.e., the i-th column vector, is the i-order vibration mode of the system.

To use the modal superposition method, the Rayleigh damping assumption is generally used, i.e., the damping matrix $[C]$ can be expressed as follows:

$$[C] = \alpha_0[M] + \alpha_1[K]$$

where,

α_0, α_1 ——the constants related to the system.

Thus, under the effects of ground motion, the equation of motion for the system can be expressed as follows:

$$[M]\{\ddot{x}(t)\} + (\alpha_0[M] + \alpha_1[K])\{\dot{x}(t)\} + [K]\{x(t)\} = -[M]\{I\}\ddot{x}_g(t) \quad (3\text{-}80)$$

By introducing Eq. (3-79) into Eq. (3-80) and pre-multiplying it with the mode shapevector $\{X\}_j^T$, we get:

$$\{X\}_j^T[M][X]\{\ddot{q}(t)\} + \{X\}_j^T(\alpha_0[M] + \alpha_1[K])[X]\{\dot{q}(t)\} + \{X\}_j^T[K][X]\{q(t)\} = -\{X\}_j^T[M]\{I\}\ddot{x}_g(t) \quad (3\text{-}81)$$

Now, consider the first term on the left side of Eq. (3-81):

$$\{X\}_j^T[M][X]\{\ddot{q}(t)\} = \{X\}_j^T[M]$$

$$\{\{X\}_1 \cdots \{X\}_i \cdots \{X\}_n\}\begin{Bmatrix}\ddot{q}_1(t)\\\ddot{q}_2(t)\\\vdots\\\ddot{q}_i(t)\\\vdots\\\ddot{q}_n(t)\end{Bmatrix}$$

$$= \{X\}_j^T[M][X]_1\ddot{q}_1(t) + \cdots + \{X\}_j^T[M][X]_i\ddot{q}_i(t) + \cdots + \{X\}_j^T[M][X]_n\ddot{q}_n(t) = \{X\}_j^T[M][X]_j\ddot{q}_j(t)$$

where the last step is derived from the orthogonality condition for the mass matrix. Similarly, using the orthogonality condition for the stiffness matrix, we can obtain the following:

$$\{X\}_j^T[K][X]\{q(t)\} = \{X\}_j^T[K]\{X\}_j q_j(t)$$
$$= \omega_j^2\{X\}_j^T[M]\{X\}_j q_j(t)$$

Similarly, the second term on the left side of Eq. (3-81) can be written as follows:

$$\{X\}_j^T(\alpha_0[M] + \alpha_1[K])[X]\{\dot{q}(t)\} = (\alpha_0 + \alpha_1\omega_j^2)\{X\}_j^T[M]\{X\}_j\dot{q}_j(t)$$

By introducing the above equations into Eq. (3-81) and dividing by $\{X\}_j^T[M]\{X\}_j$, we have:

$$\ddot{q}_j(t) + (\alpha_0 + \alpha_1\omega_j^2)\dot{q}_j + \omega_j^2 q_j(t) = -\gamma_j\ddot{x}_g(t) \quad (3\text{-}82)$$

where,

$$\gamma_j = \frac{\{X\}_j^T[M]\{I\}}{\{X\}_j^T[M]\{X\}_j} = \sum_{i=1}^n m_i X_{ji}\Big/\sum_{i=1}^n m_i X_{ji}^2 \quad (3\text{-}83)$$

Given:

$$\alpha_0 + \alpha_1\omega_j^2 = 2\xi_j\omega_j \quad (3\text{-}84)$$

Eq. (3-82) becomes:

$$\ddot{q}_j(t) + 2\xi_j\omega_j\dot{q}_j(t) + \omega_j^2 q_j(t) = -\gamma_j\ddot{x}_g(t) \quad (3\text{-}85)$$

In Eq. (3-84), ξ_j is called the mode damping ratio, which corresponds to the j-order vibration mode; furthermore, coefficients α_0 and α_1 can be determined by the frequencies and damping ratios of the first and second modes. Based on Eq. (3-84), we have:

$$\alpha_0 + \omega_1^2\alpha_1 = 2\xi_1\omega_1$$
$$\alpha_0 + \omega_2^2\alpha_1 = 2\xi_2\omega_2$$

By solving the above equations, we arrive at the following relationships:

$$\alpha_0 = \frac{2\omega_1\omega_2(\xi_1\omega_2 - \xi_2\omega_1)}{\omega_2^2 - \omega_1^2}$$
$$\alpha_1 = \frac{2(\xi_2\omega_2 - \xi_1\omega_1)}{\omega_2^2 - \omega_1^2} \quad (3\text{-}86)$$

In the above derivation, taking $j = 1, 2, \cdots, n$, successively can result in n separate differential equations, and each equation contains only one unknown quantity $q_j(t)$, which can be solved using the solution of the SDOF systems. Then, we can obtain $q_1(t), q_2(t), \cdots, q_n(t)$.

By introducing the generalized coordinates $q_j(t)(j = 1, 2, \cdots, n)$ into Eq. (3-79), we can then obtain the displacements of each particle $x_j(t)(i = 1, 2, \cdots, n)$.

If we define $\Delta_j(t)$ as the displacement of a SDOF system with a damping ratio of ξ_j and a natural frequency of ω_j, we can obtain the solution of the j-order vibration mode by comparing with Eq. (3-85), i.e., $q_j = \gamma_j\Delta_j(t)$, and the displacement of particle i is as follows:

$$x_j(t) = \sum_{j=1}^n q_j(t)X_{ji} = \sum_{j=1}^n \gamma_j\Delta_j(t)X_{ji} \quad (3\text{-}87)$$

In Eq. (3-83), γ_j is called the modal participation factor of the j-order vibration mode, which satisfies:

$$\sum_{j=1}^n \gamma_j X_{ji} = 1 \quad (3\text{-}88)$$

Let us take the 2-DOF system as an example, the primary results can be shown using the following approach.

Given:

$$x_1(t) = q_1(t)X_{11} + q_2(t)X_{21} = 1$$
$$x_2(t) = q_1(t)X_{12} + q_2(t)X_{22} = 1 \quad (3\text{-}89)$$

By multiplying the first and second equations in Eq. (3-89) by m_1X_{11} and m_2X_{22}, respectively, and adding them, we obtain the following:

$$m_1X_{11} + m_2X_{12} = q_1(t)m_1X_{11}^2 + q_2(t)m_1X_{11}\cdot X_{21} + q_1(t)m_2X_{12}^2 + q_2(t)m_2X_{12}\cdot X_{22} \quad (3\text{-}90)$$

Based on the orthogonality of the modes,

we have:
$$m_1 X_{11} X_{21} + m_2 X_{12} X_{22} = 0$$
Eq. (3-90) can then be written as follows:
$$m_1 X_{11} + m_2 X_{12} = q_1(t)(m_1 X_{11}^2 + m_2 X_{12}^2)$$
This equation can be rewritten as follows:
$$q_1(t) = \frac{m_1 X_{11} + m_2 X_{12}}{m_1 X_{11}^2 + m_2 X_{12}^2} = \gamma_1$$

Similarly, we can obtain:
$$q_2(t) = \gamma_2$$
Thus, Eq. (3-89) becomes:
$$\gamma_1 X_{11} + \gamma_2 X_{21} = 1$$
$$\gamma_1 X_{12} + \gamma_2 X_{22} = 1$$
The above equation depicts the particular case of Eq. (3-88) for 2-DOF systems.

3.8 Horizontal Seismic Effect and Response of Multi-Degree-of-Freedom Systems

Engineers care about the maximum value of particle responses. The maximum displacement and absolute acceleration of each particle can be obtained using the modal superposition method presented above; however, the modal superposition method is slightly complicated and inconvenient for practical engineering applications. Based on the modal analysis method of MDOF systems and response spectrum analysis of SDOF systems, a response spectrum method of MDOF systems is commonly used. In addition, a simple and practical base shear method is also applied for particular structures. Also, the time history method is often employed for important and complex structures.

3.8.1 Response Spectrum Method

The horizontal seismic effect of MDOF systems can be represented by the inertial force of each particle. Furthermore, the seismic effect of a particle i is as follows:
$$F_i(t) = -m_i [\ddot{x}_g(t) + \ddot{x}_i(t)]$$
Based on Eq. (3-88), $\ddot{x}_g(t)$ can be written as follows:
$$\ddot{x}_g(t) = \ddot{x}_g(t) \sum_{j=1}^{n} \gamma_j X_{ji} = \sum_{j=1}^{n} \ddot{x}_g(t) \gamma_j X_{ji}$$
Based on Eq. (3-87), we have the following:
$$\ddot{x}_i(t) = \sum_{j=1}^{n} \gamma_j \ddot{\Delta}_j(t) X_{ji} \quad (3-91)$$
Then:

$$F_i(t) = -m_i \sum_{j=1}^{n} \gamma_j X_{ji} [\ddot{x}_g(t) + \ddot{\Delta}_j(t)]$$
$$= \sum_{j=1}^{n} F_{ji}(t) \quad (3-92)$$
where,
$$F_{ji}(t) = -m_i \gamma_j X_{ji} [\ddot{x}_g(t) + \ddot{\Delta}_j(t)] \quad (3-93)$$
Which is called the horizontal seismic effect of particle i corresponding to the j-order vibration mode.

We can determine the variation in the horizontal seismic effect of particle i as a function of time using Eq. (3-92), and the maximum value can then be determined. However, an even simpler approach is to use the response spectrum of SDOF systems and Eq. (3-93) to obtain the maximum horizontal seismic effect and the response (such as the moment, shear force, axial force, displacement, etc.) of each particle corresponding to the j-order vibration mode. Moreover, the response of MDOF systems resulting from the horizontal seismic effect can be determined after combining the response corresponding to each vibration mode.

Based on Eq. (3-93), we have:
$$F_{ji_{\max}} = \left| \frac{\ddot{x}_g(t) + \ddot{\Delta}_j(t)}{g} \right|_{\max} G_i \gamma_j X_{ji}$$
$$(3-94)$$
Given:
$$\alpha_j = \left| \frac{\ddot{x}_g(t) + \ddot{\Delta}_j(t)}{g} \right|_{\max} \quad (3-95)$$
where,
α_j——the seismic effect coefficient of the

j-th SDOF systems with a period $T_j = \dfrac{2\pi}{\omega_j}$ in the j-order vibration mode. This value can be determined by the seismic response coefficient of the SDOF systems.

Thus, the maximum seismic effect of the i-th particle in the j-th mode is as follows:

$$F_{ji_{max}} = \alpha_j \gamma_j X_{ji} G_i \qquad (3\text{-}96)$$

Based on Eq. (3-96) and the α–T curves given by the **Code for Seimic Design of Buildings** (GB 50011—2010) Version 2016, it is convenient to calculate the maximum seismic effect of a certain mode. Furthermore, we can obtain the effects (i.e., the moment, shear force, axial force, displacement, etc.) of the earthquake corresponding to each mode using structural mechanics.

Based on Eq. (3-92), the seismic effect of a structure at any time is equal to the sum of the seismic effect under each mode. The seismic effect of each mode may not reach the peak value at the same time. Therefore, the maximum seismic effect is not simply the sum of the maximum values of each mode. Thus, we should consider a method to combine the different maximum values of each mode.

Modal maxima do not occur simultaneously, so any combination of modal maxima may lead to results that are either conservative or unconservative. The accuracy of results depends on which modal combination technique is used and the dynamic properties of the system being analyzed. Three of the most commonly used modal combination methods are:

1) Sum of absolute values

This method assumes that the maximum modal values occur simultaneously, therefore the maximum absolute values of each mode are summed as the maximum values of the approximate combination values, which leads to very conservative results. The response of any given degree-of-freedom of the system is estimated as:

$$S_{i_{max}} \approx \sum_{n=1}^{L} |S_{i,n_{max}}| \qquad (3\text{-}97)$$

2) Square root of the sum of the squares (SRSS)

The SRSS method assumes that the individual modal maxima are statistically independent (uncorrelated). Based on the stochastic vibration theory, if the movement of the floor is considered as a stationary random process, we can use the "square root of the sum of the squares" method to determine the seismic effect as Eq. (3-98). The SRSS method generally leads to values that are closer to the "exact" values than those obtained using the sum of the absolute values. The results of a SRSS analysis can be significantly unconservative if the modal periods are closely spaced.

$$S = \sqrt{\sum_{j=1}^{n} S_j^2} \qquad (3\text{-}98)$$

where,

S——the sum of the seismic effect at a certain point in a structure;

S_j——the corresponding seismic effect at the j-th mode.

When using the modal response spectrum method, we cannot use the "square root of the sum of the squares" method to combine all of the modes; instead, the total effect of the earthquake is calculated first, and then the seismic response can be obtained.

3) Complete quadratic combination (CQC)

The method is based on the stochastic process theory (correlated), which takes into account not only the square terms of each main mode shape but also the coupling terms. For more complex structures, such as structures with a plane and torsional coupling, the results obtained by using the CQC method are more accurate. This method has been integrated into several commercial analysis programs.

$$S_{i_{max}} \approx \sqrt{\sum_{n=1}^{L}\sum_{m=1}^{L} S_{i,n_{max}} \rho_{n,m} S_{i,m_{max}}}$$

$$(3\text{-}99)$$

[**Example 3-3**]

A three-storey frame structure is depicted in Fig. 3-14. We can assume that the stiffness of the beam is infinite. The columns have the same sections, and the mass of each storey and the three vibration modes, as well as their periods, are depicted in Fig. 3-14. The seismic intensity for the design is 7. Furthermore, the structure is located at a Class I site. $T_g = 0.3$ s, and the damping ratio ξ of the structure is 0.05. Determine the moment of the frame beams under a horizontal seismic effect using the modal response spectrum method.

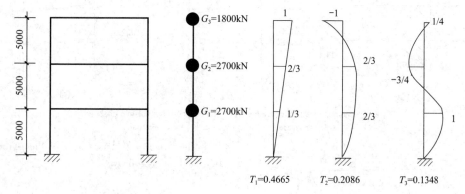

Fig. 3-14 Mass of each storey and vibration modes of the three-storey frame

[**Solution**]:

Look up Table 4-5 and the seismic response coefficient curve (Fig. 3-9).

For the first mode:
$$T_1 = 0.4665 \text{ s} > T_g$$

Seismic response coefficient:
$$\alpha_1 = \left(\frac{T_g}{T_1}\right)^{0.9} \alpha_{max} = \left(\frac{0.3}{0.4665}\right)^{0.9} \times 0.08 = 0.054$$

Vibration mode participation coefficient:
$$\gamma_1 = \frac{2700 \times \frac{1}{3} + 2700 \times \frac{2}{3} + 1800 \times 1}{2700 \times \left(\frac{1}{3}\right)^2 + 2700 \times \left(\frac{2}{3}\right)^2 + 1800 \times 1^2} = 1.364$$

Seismic effect:
$$F_{11} = 0.054 \times 1.364 \times \frac{1}{3} \times 2700 = 66.3 \text{ kN}$$

$$F_{12} = 0.054 \times 1.364 \times \frac{2}{3} \times 2700 = 132.6 \text{ kN}$$

$$F_{13} = 0.054 \times 1.364 \times 1 \times 1800 = 132.6 \text{ kN}$$

For the second mode:
$T_2 = 0.2086$ s, $0.1 < T_2 < T_g$, then $\alpha_2 = \alpha_{max} = 0.08$

$$\gamma_2 = \frac{2700 \times \frac{2}{3} + 2700 \times \frac{2}{3} + 1800 \times (-1)}{2700 \times \left(\frac{2}{3}\right)^2 + 2700 \times \left(\frac{2}{3}\right)^2 + 1800 \times (-1)^2} = 0.429$$

$$F_{21} = 0.08 \times 0.429 \times \frac{2}{3} \times 2700 = 61.8 \text{ kN}$$

$$F_{22} = 0.08 \times 0.429 \times \frac{2}{3} \times 2700 = 61.8 \text{ kN}$$

$$F_{23} = 0.08 \times 0.429 \times (-1) \times 1800 = -61.8 \text{ kN}$$

For the third mode:
$T_3 = 0.1348$ s, $0.1 < T_3 < T_g$, then $\alpha_3 = \alpha_{max} = 0.08$

$$\gamma_3 = \frac{2700 \times 1 + 2700 \times \left(-\frac{3}{4}\right) + 1800 \times \frac{1}{4}}{2700 \times 1^2 + 2700 \times \left(-\frac{3}{4}\right)^2 + 1800 \times \left(\frac{1}{4}\right)^2} = 0.26$$

$$F_{31} = 0.08 \times 0.26 \times 1 \times 2700 = 56.2 \text{ kN}$$

$$F_{32} = 0.08 \times 0.26 \times \left(-\frac{3}{4}\right) \times 2700 = -42.1 \text{ kN}$$

$$F_{33} = 0.08 \times 0.26 \times \frac{1}{4} \times 2700 = 9.4 \text{ kN}$$

Based on the seismic effect of each vibration mode above, the moment of the frame under earthquakes corresponding to each mode can be calculated, which is illustrated in Fig. 3-15.

Based on the combination principle of the "square root of the sum of the squares" method, the moment diagram of the frame can be obtained by combining all the modes, which is shown in Fig. 3-16.

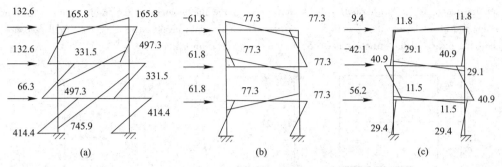

Fig. 3-15 Moment diagram of each vibration mode under the seismic effect (kN)
(a) The first mode; (b) The second mode; (c) The third mode

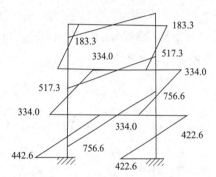

Fig. 3-16 Combined moment diagram of the frame (kN)

3.8.2 Base Shear Method

For general structures, we can use the modal response spectrum method to calculate the seismic response. However, if the buildings meet the following conditions, the base shear method can be used.

(1) The building height is less than 40 m;

(2) The mass and stiffness are uniformly distributed with the building height;

(3) The shear deformation is predominant under earthquakes;

(4) The torsional effect under earthquakes could be neglected.

If a structure meets the above conditions, its reaction to earthquakes is typically dominated by the first mode and is nearly linear.

Based on the modal response spectrum method, the seismic action of the modal particle is as follows:

$$F_{ji_{max}} = \alpha_j \gamma_j X_{ji} G_i$$

The structure's base shear of the j-vibration mode is as follows:

$$S_{jE} = \sum_{i=1}^{n} \alpha_j \gamma_j X_{ji} G_i = \alpha_1 G_E \sum_{i=1}^{n} \frac{\alpha_j}{\alpha_1} \gamma_j X_{ji} \frac{G_i}{G_E} \quad (3\text{-}100)$$

where,

α_1——the seismic effect coefficient of the first mode;

G_E——the representative value of gravity load, $G_E = \sum_{i=1}^{n} G_i$, where G_i is the representative value of the gravity load for particle i.

Based on the modal combination principle of the "square root of the sum of square" (SRSS) method, the base shear of a structure is:

$$F_{Ek} = S = \sqrt{\sum_{j=1}^{n} S_{jE}^2}$$
$$= \alpha_1 G_E \sqrt{\sum_{j=1}^{n} \left[\sum_{i=1}^{n} \frac{\alpha_j}{\alpha_i} \gamma_j X_{ji} \frac{G_i}{G_E} \right]^2}$$
$$= \alpha_1 G_E \zeta \quad (3\text{-}101)$$

where ζ is a coefficient given by:

$$\zeta = \sqrt{\sum_{j=1}^{n} \left[\sum_{i=1}^{n} \frac{\alpha_j}{\alpha_i} \gamma_j X_{ji} \frac{G_i}{G_E} \right]^2}$$

Clearly, when $n=1$, the coefficient $\zeta = 1$. Additionally, based on statistical data, when $n > 1$, the coefficient ζ is influenced by the modes as well as the number of mass particles. When $n=3$ or 4, the coefficient ζ is approximately 0.85. Lastly, when $n>4$, the coefficient ζ is approximately 0.8.

To simplify the calculations, the Chinese ***Code for Seismic Design of Buildings*** (GB 50011—2010) Version 2016 specifies that when $n=1$, the coefficient ζ becomes 1; when $n>1$, the coefficient ζ is 0.85. Additionally, the equivalent representative value of gravity load can be defined as $G_{eq} = \zeta G_E$. Consequently, when we calculate the seismic effect using the base shear method, the base shear force (or total horizontal seismic action) is as follows:

$$F_{Ek} = \alpha_1 G_{eq} \quad (3\text{-}102)$$

Based on the assumption that the structural response is dominated by the first mode, the horizontal seismic effect of each particle is approximately equal to the effect corresponding to the first mode, namely:

$$F_i = F_{1i} = \alpha_1 \gamma_1 X_{1i} G_i \quad (3\text{-}103)$$

Then, according to the assumption that the shape of the first mode is close to linear, we have:

$$X_{1i} = \eta H_i \quad (3\text{-}104)$$

where,

H_i——the calculated height of particle i.

By introducing Eq. (3-104) into Eq. (3-103), we can obtain the following:

$$F_i = \alpha_1 \gamma_1 \eta H_i G_i \quad (3\text{-}105)$$

Then, based on the condition that the sum of the horizontal seismic effect of each particle is equal to the total horizontal seismic effect, we have:

$$F_{Ek} = \sum_{k=1}^{n} F_k = \sum_{k=1}^{n} \alpha_1 \gamma_1 \eta H_k G_k$$

$$\alpha_1 \gamma_1 \eta = \frac{F_{Ek}}{\sum_{k=1}^{n} H_k G_k}$$

By introducing the formulas above into Eq. (3-105), we can obtain the following:

$$F_i = \frac{H_i G_i}{\sum_{k=1}^{n} H_k G_k} \cdot F_{Ek} \quad (3\text{-}106)$$

The horizontal seismic action calculated using Eq. (3-104) can suitably represent the seismic action of structures with high stiffness, such as masonry structures. However, when the fundamental period of a structure is relatively long and the characteristic period of the site T_g is short, the seismic action at the top of a structure calculated using Eq. (3-104) is relatively small. Thus, the **Code for Seismic Design of Buildings** (GB 50011—2010) Version 2016 prescribes that an additional horizontal seismic effect should be added at the top of the building for $T_1 > 1.4 T_g$, namely:

$$\Delta F_n = \delta_n F_{Ek}$$

Then, the remaining horizontal seismic effect based on Eq. (3-106) can be distributed; hence, considering the correction, the horizontal seismic effects of the top can be represented as follows:

$$F_n = \frac{H_n G_n}{\sum_{k=1}^{n} H_k G_k}(1 - \delta_n) F_{Ek} + \delta_n F_{Ek}$$

$$(3\text{-}107)$$

The others can be represented as follows:

$$F_i = \frac{H_i G_i}{\sum_{k=1}^{n} H_k G_k}(1 - \delta_n) F_{Ek} \quad (3\text{-}108)$$

where,

δ_n——the coefficient of the additional seismic effect.

For multi-storey reinforced concrete and steel buildings, δ_n can be determined by the characteristic period of the site T_g and the basic period of the structure T_1 using Table 4-7 in Chapter 4.

[**Example 3-4**]

Determine the moment diagram of the three-storey frame in [Example 3-3] using the base shear method.

[**Solution**]:

The representative value of the structure gravity force:

$$G_{eq} = 0.85 \times (1800 + 2700 \times 2)$$
$$= 6120 \text{ kN}$$

The sum of the horizontal effects:

$$F_{Ek} = \alpha_1 G_{eq} = 0.054 \times 6120 = 330.48 \text{ kN}$$

Because $T_1 = 0.4665$ s $> 1.4 T_g = 1.4 \times 0.3 = 0.42$ s, the coefficients of the additional seismic effects can be obtained from Table 4-7:

$$\delta_n = 0.08 T_1 + 0.07$$
$$= 0.08 \times 0.4665 + 0.07 = 0.1073$$

$$\sum_{k=1}^{n} H_k G_k = 2700 \times 5 + 2700 \times 10 + 1800 \times 15$$
$$= 67,500$$

$$F_1 = \frac{2700 \times 5}{67{,}500} \times (1 - 0.1073) \times 330.48$$
$$= 59.00 \text{ kN}$$
$$F_2 = \frac{2700 \times 10}{67{,}500} \times (1 - 0.1073) \times 330.48$$
$$= 118.01 \text{ kN}$$
$$F_3 = \left[\frac{1800 \times 15}{67{,}500} \times (1 - 0.1073) + 0.1073\right] \times$$
$$330.48 = 153.47 \text{ kN}$$

The moment diagram of the frame caused by seismic effects is illustrated in Fig. 3-17.

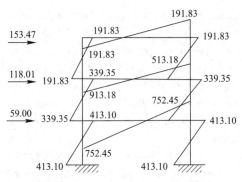

Fig. 3-17 Moment diagram of the frame (kN)

3.9　Time History Method

As described above, for general structures, the modal response spectrum method can be used to calculate the seismic effect. However, for particularly irregular buildings, i.e., important buildings and buildings with considerable height and fortification intensity, the time history analysis method should be used to supplement the calculation to ensure the safety of these buildings in the presence of earthquakes. Furthermore, when the elastoplastic analysis is conducted, the structure has already entered a nonlinear phase; therefore, the modal response spectrum method is no longer applicable, and the elastoplastic time history analysis should be used instead.

When the time history analysis of a structure in the presence of earthquakes is needed, a numerical calculation method is generally used, such as the Wilson-θ method and the introduced linear acceleration method.

3.9.1　Linear Acceleration Method for Multi-Degree-of-Freedom Systems

The equation of motion for multi-degree-of-freedom systems under the action of ground motion is as follows:
$$[M]\{\ddot{x}(t)\} + [C]\{\dot{x}(t)\} + [K]\{x(t)\} = -[M]\{I\}\ddot{x}_g(t) \quad (3\text{-}109)$$

The incremental form of the equation of motion is given by subtracting the equation of motion at t_{k-1} from the equation of motion at t_k, which is:
$$[M]\{\Delta\ddot{x}\} + [C]\{\Delta\dot{x}\} + [K]\{\Delta x\} = -[M]\{I\}\Delta\ddot{x}_g$$

Similar to the derivation in Section 3.5.3, the result can be written as follows:
$$[K^*]\{\Delta x\} = \{\Delta P^*\} \quad (3\text{-}110)$$

where, $[K^*] = [K] + \frac{3}{\Delta t}[C] + \frac{6}{\Delta t^2}[M]$

$$\{\Delta P^*\} = -[M]\{I\}\Delta\ddot{x}_g + [M]\left(\frac{6}{\Delta t}\{\dot{x}_{k-1}\} + 3\{\ddot{x}_{k-1}\}\right) + [C]\left(3\{\dot{x}_{k-1}\} + \frac{\Delta t}{2}\{\ddot{x}_{k-1}\}\right)$$
$$(3\text{-}111)$$

where $\{\Delta x\}$ is given by Eq. (3-111); $\{\Delta\dot{x}\}$ and $\{\Delta\ddot{x}\}$ are:
$$\{\Delta\dot{x}\} = \frac{3}{\Delta t}\{\Delta x\} - 3\{\dot{x}_{k-1}\} - \frac{\Delta t}{2}\{\ddot{x}_{k-1}\}$$
$$\{\Delta\ddot{x}\} = \frac{6}{\Delta t^2}\{\Delta x\} - \frac{6}{\Delta t}\{\dot{x}_{k-1}\} - 3\{\ddot{x}_{k-1}\}$$
$$(3\text{-}112)$$

$$\{x_k\} = \{x_{k-1}\} + \{\Delta x\}$$
and $\{\dot{x}_k\} = \{\dot{x}_{k-1}\} + \{\Delta\dot{x}\}$
$$= \frac{3}{\Delta t}\{\Delta t\} - 2\{\dot{x}_{k-1}\} - \frac{\Delta t}{2}\{\ddot{x}_{k-1}\}$$
$$(3\text{-}113)$$

$$\{\ddot{x}_k\} = \{\ddot{x}_{k-1}\} + \{\Delta\ddot{x}\}$$
$$= \frac{6}{\Delta t^2}\{\Delta x\} - \frac{6}{\Delta t}\{\dot{x}_{k-1}\} - 2\{\ddot{x}_{k-1}\}$$

By repeating the above steps, the displacement, velocity and acceleration of the system at each $t_k (k = 1, 2, \cdots, n)$ can be calculated step-by-step based on the initial conditions of the system.

3.9.2 Wilson-θ Method for Multi-Degree-of-Freedom Systems

A basic requirement of numerical calculation methods is the convergence of the algorithm. As introduced in Section 3.9.1, when the natural vibration period of the system is short and the calculation step is large, the linear acceleration method may exhibit computational divergence, which indicates that the calculated data becomes increasingly large until it overflows (which cannot be represented by a computer). For multi-degree-of-freedom systems, the smallest natural vibration period may be small, and the calculation step Δt must be small enough to ensure that the calculation converges. Instead, the linear acceleration method is a conditional convergence algorithm.

The Wilson-θ method is a numerical method of unconditional convergence, which is an improved method based on the linear acceleration method. As long as the parameter θ is appropriate, a convergent result can be obtained regardless of the magnitude of Δt. Certainly, a higher value of Δt leads to larger errors.

At time $t + \theta \Delta t$, the equation of motion for multi-degree-of-freedom systems can be represented by the following:

$$[M]\{\ddot{x}_{t+\theta\Delta t}(t)\} + [C]\{\dot{x}_{t+\theta\Delta t}(t)\}$$
$$+ [K]\{x_{t+\theta\Delta t}(t)\} = -[M]\{I\}$$
$$[\ddot{x}_{gt} + \theta(\ddot{x}_{gt+\Delta t} - \ddot{x}_{gt})]$$

The Wilson-θ method assumes that the system acceleration response has a linear variation during the period $(t, t+\theta \Delta t)$. Using the linear acceleration method during $(t, t+\theta \Delta t)$, the displacement of the system $\{x\}_{t+\Delta t}$ at time $t+\theta \Delta t$ can be obtained when the time step is $\theta \Delta t$. Using Eq. (3-113), the acceleration at time $t + \theta \Delta t$ is:

$$\{\ddot{x}\}_{t+\theta\Delta t} = \frac{6}{(\theta \Delta t)^2}(\{x\}_{t+\theta\Delta t} - \{x\}_t)$$
$$- \frac{6\{\dot{x}\}_t}{\theta \Delta t} - 2\{\ddot{x}\}_t$$

Based on the assumption that the acceleration changes linearly, the acceleration at $t + \Delta t$ can be obtained via interpolation to get:

$$\{\ddot{x}\}_{t+\Delta t} = \{\ddot{x}\}_t + \frac{1}{\theta}(\{\ddot{x}\}_{t+\theta\Delta t} - \{\ddot{x}\}_t) = \{\ddot{x}\}_t$$
$$+ \frac{1}{\theta}\left(\frac{6}{(\theta\Delta t)^2}(\{x\}_{t+\theta\Delta t} - \{x\}_t) - \frac{6}{\theta\Delta t}\{\dot{x}\}_t - 3\{\ddot{x}\}_t\right)$$
$$= \frac{6}{\theta(\theta\Delta t)^2}(\{x\}_{t+\theta\Delta t} - \{x\}_t) - \frac{6}{\theta^2 \Delta t}\{\dot{x}\}_t$$
$$+ \left(1 - \frac{3}{\theta}\right)\{\ddot{x}\}_t$$

Based on $\{\ddot{x}\}_{t+\Delta t}$ and the basic linear interpolation formula as well as Eq. (3-18) and Eq. (3-19), we can obtain the following:

$$\{\dot{x}\}_{t+\Delta t} = \{\dot{x}\}_t + \frac{\Delta t}{2}(\{\ddot{x}\}_{t+\Delta t} + \{\ddot{x}\}_t)$$

(3-114)

$$\{x\}_{t+\Delta t} = \{x\}_t + \Delta t\{\dot{x}\}_t + \frac{1}{3}\Delta t^2 + \frac{\Delta t^2}{6}\{\ddot{x}\}_{t+\Delta t}$$

(3-115)

3.9.3 Seismic Response Calculation for Multi-Degree-of-Freedom Nonlinear Systems

For linear systems with multiple degrees of freedom, the quasi-stiffness matrix in Eq. (3-110) is a constant matrix; however, for nonlinear systems, although we assume that the variation in the stiffness of the members in the system can be neglected during Δt, the stiffness of the members changes at different stages. Thus, the stiffness matrix and the quasi-stiffness matrix of the system are continuously changing. Therefore, the stiffness matrix must be considered when calculating the nonlinear seismic response of the system.

For the inter-laminar shear model for multi-degree-of-freedom systems, a restoring force model between each layer must be established to analyze the nonlinear seismic response. When the linear acceleration method is used for time-history analysis, each interlamination needs to be distinguished in each step to determine whether the inter-laminar stiffness coefficient K_i ($i = 1, 2, \cdots, n$) must be modified. When a stiffness coefficient is modified, regenerating the stiffness matrix and the quasi-stiffness matrix is necessary before proceeding.

Chapter 4
Earthquake Effect and Seismic Design Principles

4.1 Building Classification and Seismic Fortification

4.1.1 Seismic Fortification Classification

In seismic design, buildings are usually classified into different seismic fortification categories based on their architectural function and functional importance (Fig. 4-1). Various seismic fortification criteria have been established for the different seismic fortification categories.

Fig. 4-1 Building classification and seismic fortification

The Chinese *General Code for Seismic Precaution of Buildings and Municipal Engineering* (GB 55002—2021) classifies buildings are into four categories based on their architectural importance, societal and economic impact after damage and role in earthquake events.

(1) Particular protection buildings and municipal engineering (Category A) are extremely important or those whose failure could result in serious secondary disasters;

(2) Emphasized protection buildings and municipal engineering (Category B) are those for which the continual function during earthquakes is necessary or that should be rapidly recoverd after earthquakes;

(3) Standard protection buildings and municipal engineering (Category C) are those not classified as either Category A, B, or D;

(4) Moderate protection buildings and municipal engineering (Category D) are considered less important, which is only used for a small number of people and its failure would not cause secondary disasters.

4.1.2 Seismic Fortification Criterion

The seismic fortification criterion for each seismic fortification category is as follows:

(1) For Category C, the seismic measures and earthquake action should be determined according to the local seismic intensity.

(2) For Category B, the seismic measures and earthquake action should be determined based on the local seismic intensity. When the seismic intensity is 6—8, the seismic measures should generally be taken in accordance with the intensity one degree higher than the local seismic intensity. However, when the local seismic intensity is 9, the seismic measures should be taken in accordance with intensity larger than 9. Furthermore, the seismic measures for the foundation should comply with the corresponding provisions.

(3) For Category A, the earthquake action should be determined according to the approved seismic safety evaluation report and an intensity higher than that of the local seismic intensity. Furthermore, the seismic measures should be taken in accordance with the intensity one degree higher than the local seismic intensity when the local seismic intensity is 6—8. However, when the local seismic intensity is 9, the seismic measures should be taken in accordance with an intensity exceeding 9.

(4) For Category D, the seismic measures could be taken in accordance with the intensity lower than the local seismic intensity, but it is

not allowed in zones with seismic intensity 6. Generally, the earthquake action is determined according to the local seismic intensity.

(5) When the site category is Class I, it is allowed to take seismic structural measures according to the requirements of the local seismic intensity for Category A or B. For Category C, the seismic measures should be taken in accordance with the intensity one degree lower than the local seismic intensity, but not lower than 6.

(6) For urban bridges, the minor earthquake action should be adjusted by multiplying the corresponding importance coefficient in accordance with the different seismic fortification categories. The importance coefficient of urban bridges of Categories A, B, C and D should not be less than 2.0, 1.7, 1.3 and 1.0, respectively.

If the buildings in Categories B, C and D are located in a seismic intensity 6 zone, earthquake action measures are not required.

In summary, the seismic fortification criterion associated with each seismic fortification category is listed in Table 4-1.

Seismic fortification category and seismic fortification criterion   Table 4-1

Protection category	Category D: Moderate protection buildings	Category C: Standard protection buildings	Category B: Emphasized protection buildings	Category A: Particular protection buildings
Seismic force (seismic effect)	Equal to the seismic intensity	Equal to the seismic intensity	Equal to the seismic intensity	Higher than the seismic intensity or based on the site safety evaluation
Structural details (seismic measures)	Appropriately lower than the seismic intensity except for intensity Ⅵ(6)	Equal to the seismic intensity	One degree higher than the seismic intensity	One degree higher than the seismic intensity

4.1.3 Seismic Fortification Objective

An earthquake is a stochastic event and the influencing factors include not only the specific time and place, but also the intensity and frequency of the event. Thus, an objective that requires a structure to remain undamaged against every possible earthquake is neither economic nor scientific. A more reasonable seismic fortification objective is that a structure should have different resistance levels for different intensities and frequencies during its lifetime service. Particularly, for frequently occurring earthquakes, the structure is required to not be damaged. For such small earthquakes, the objective is achievable both technically and economically. For rarely occurring earthquakes, due to their high intensity and low probability, moderate damage is acceptable; however, the structure should not collapse under any circumstances.

The Chinese national *Code for Seismic Design of Buildings* (GB 50011—2010) Version 2016 proposes a "two-stage and three-level" seismic design method (Table 4-2).

The "three-level" can be specified as follows:

Level 1 (minor earthquake level): When subjected to frequently occurring earthquakes for which the intensity is lower than the local seismic intensity, the buildings will either not be damaged or be slightly damaged with continued service capability without repair, i.e., no damage under minor earthquakes.

Level 2 (moderate earthquake level): When subjected to earthquakes for which the intensity is equal to the local seismic intensity, the buildings will be damaged with continued service capability after ordinary repair or without repair, i.e., repairable under moderate earthquakes.

Level 3 (major earthquake level): When subjected to rarely occurring earthquakes for which the intensity is expected to be higher than the local seismic intensity, the buildings will neither collapse nor suffer damage that would endanger human lives, i.e., no collapse under major earthquakes.

The "two-stage" can be specified as follows:

Stage 1: The elastic force should be examined under the basic load combination considering minor earthquake action and other load effects. The elastic deformation should be examined under minor earthquake conditions (corresponding to Level 1).

Stage 2: The elastoplastic deformation should be examined under major earthquake conditions (corresponding to Level 3).

For Level 2, the seismic conceptual design and seismic detailing are used as a guarantee.

It is important to note that the fortification objectives are determined by a country's current scientific and economic conditions.

Two-stage and three-level seismic design method   Table 4-2

Three-level	Level 1: No damage under minor earthquakes	Level 2: Repairable damage under moderate earthquakes	Level 3: No collapse under major earthquakes
Two-stage	Examine: Elastic force, elastic deformation (Stage 1)	Seismic details (seismic measures)	Examine: Elastoplastic deformation (Stage 2)

4.1.4 Three Levels

Here, minor earthquakes are referred to as frequently occurring earthquakes, moderate earthquakes are referred to as local seismic intensity earthquakes, and major earthquakes are referred to as rarely occurring earthquakes.

In China, it is assumed that the Probability Density Function (PDF) of earthquake intensity follows a Limit 3 distribution as follows:

$$f(x) = \frac{k(\omega - I)^{k-1}}{(\omega - \varepsilon)} e^{-\left(\frac{\omega - I}{\omega - \varepsilon}\right)^k} \quad (4-1)$$

Its distribution function is:

$$F(I) = e^{-\left(\frac{\omega - I}{\omega - \varepsilon}\right)^k} \quad (4-2)$$

where,

ω——the upper limit of the seismic intensity set to $\omega = 12$;

ε——the mode intensity, which is the intensity corresponding to the peak value in the PDF curve determined by the statistics during the design period in the earthquake zone;

I——the seismic intensity;

k——the shape coefficient.

Minor earthquakes (frequently occurring earthquakes) occur with the highest frequency; thus, their intensity is equal to the mode intensity. The probability of having an intensity that is less than the mode intensity can be calculated as follows:

$$I = \varepsilon \Rightarrow F(I) = e^{-1} = 36.8\%$$

Thus, the probability beyond the mode intensity is:

$$1 - F(I) = 63.2\%$$

Major earthquakes will result in catastrophic loss; thus, it must be an event of small probability.

The probability of a major earthquake exceeding the design service life of a structure is approximately 2%—3%, according to the ***Code for Seismic Design of Buildings*** (GB 50011—2010) Version 2016.

A moderate earthquake is similar to a local seismic intensity earthquake. The intensity values of a moderate earthquake can usually be found in the "China Seismic Ground Motion Parameter Zonation Map" or the ***Code for Seismic Design of Buildings*** (GB 50011—2010) Version 2016.

Based on the survey and intensity probability study of 45 cities and towns spread across northern, southwestern and northwestern China, a moderate earthquake is approximately an earthquake whose exceedance probability is 10% during the design service life. After calculating the average deviation between the mode intensity and the intensity with an exceedance probability of 10% during the design service life of the 45 cities and towns, the result is 1.55. For major earthquakes, the intensity is approximately one level higher than that of moderate earthquakes. (Fig. 4-2). In fact, for moderate intensities 6, 7, 8 and 9, the major intensity is relatively 7, 8, 9 and 9+, respectively.

The return period of an earthquake is often used to reflect the intensity. For example, for general industrial and civil buildings, the peak ground acceleration with an exceedance probability of 10% for 50 years is taken as the design ground acceleration. Thus, a structure has a 90% chance of not being overloaded during the 50 years of service life, i.e., the exceedance probability of one year is 2.10×10^{-3}, and the return period is 475 years. A table showing the relationship between the design service life, exceedance probability and return period is provided in Table 4-3.

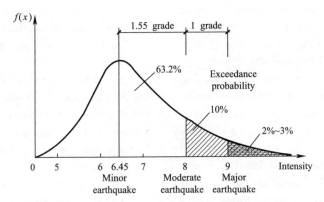

Fig. 4-2 Relationship between the three earthquake levels

Relationship between the design service life, exceedance probability and return period

Table 4-3

Exceedance Probability \ Design service life (year)	10	20	30	40	50	100
2%	495	990	1485	1980	2475	4950
3%	328	657	985	1313	1642	3283
10%	95	190	285	390	475	950
20%	45	90	135	180	225	449
30%	29	57	84	113	140	281
40%	20	40	59	79	98	196
50%	15	29	44	58	72	145
60%	11	22	33	44	55	110
63.2%	10	20	30	40	50	60

4.2　Seismic Conceptual Design

In the 1970s, the concept of "seismic conceptual design" was first proposed, which is more important than parameter-based design. Because an earthquake is a stochastic process, the seismic action to which a building is subjected during its lifetime cannot be accurately quantified. Additionally, a building is not a homogeneous block but instead a complicated assemblage of parts. Y. L. Xu said about the conceptual seismic design: "There are three levels of impact on structural safety: the impact of structural schemes can be high in percentage, the impact of internal force analysis can be up to a few percent, and the cross-section design deviation is only a few percent." For structural analysis, significant simplifications are always needed, which is both a disadvantage and a limitation in this procedure. A building with good seismic performance can hardly be obtained solely by "accurate" calculations. Seismic measures are used to compensate for this disadvantage and improve the seismic performance of buildings; however, it does not solve the problem. From lessons learned from previous earthquake disasters and engineering experience, it has been realized that seismic design should be considered in the early stages of building design. The seismic design at this stage is usually referred to as seismic conceptual design. It primarily includes the following five aspects.

1. Urban Planning

In terms of urban planning, the emphasis should be on preventing or limiting secondary disasters. In addition, the protection of service functions in urban areas after earthquakes should be considered. For example, during the Kanto Earthquake in 1923, approximately 130,000 buildings were damaged due to earthquake actions according to statistics. However, the earthquake occurred during cooking time in the afternoon, which resulted in numerous fires around the city. Because most water pipes were damaged and traffic was blocked, the fire went out of control and burned down more than 450,000 buildings, three times more than the damage caused by the earthquake itself.

2. Construction Sites

(1) An overall assessment is carried out on the location of the site which is favorable, unfavorable and hazardous areas for earthquake resistance. It is required to avoid unfavorable sites, and in case they cannot be avoided, effective measures should be taken; Categories A, B and C buildings should not be constructed in hazardous areas.

(2) When the construction site is Class I, the seismic resistance measures of Categories A and B are based on the intensity of fortification, Category C construction is allowed to be reduced by one degree; when it is 6 degrees, it should not be reduced.

(3) Foundation requirements:

(a) The foundation of a structure should not be set on different foundations;

(b) The foundation of a structure should not be partially set on natural foundation and pile foundations;

(c) If the foundation consists of soft cohesive soil, liquefied soil, new fill or serious non-uniformity, the uneven settlement of the foundation and other adverse effects should be estimated during seismic design and appropriate measures should be taken accordingly.

3. Configuration Regularity

Earthquake damage analysis indicates that a building with a simple and symmetrical configuration has a better performance in earthquake conditions. The reason is that in these types of buildings, the analysis better reflects the actual seismic response and it is easier to perform seismic measures. However, from the perspective of architectural art, if every building looks the same, it would be monotonous. Thus, the concept of "regularity" is proposed, which includes plan and vertical structural configurations, lateral-force-resisting members, and mass, stiffness and strength distributions in the horizontal and vertical directions.

(1) The vertical configuration regularity should predominate in elevation, both in geometry and variation of storey stiffness and strength, in order to avoid the appearance of a soft or weak storey;

(2) The vertical configuration should include uniformity and continuity, avoiding drastic changes;

(3) The shape of a rectangle, trapezoid or triangle without abrupt change is preferred;

(4) In the Chinese *Code for Seismic Design of Buildings* (GB 50011—2010) Version 2016, some critical vertical irregularities are defined and quantified. The first relates to soft and weak storeys, a soft storey shows a significant decrease in lateral stiffness from that immediately above and a weak storey is one in which there is a significant reduction in strength compared to that above. A pure soft storey problem exists when the lateral stiffness of one storey (most often the first storey) is significantly less than that of adjacent upper storeys. The soft storey displaces substantially more than the storeys on top of it, aggravating the P-D loads and leading to instability and failure of the soft storey. A pure weak storey problem exists when the strength of the weak storey (most often the first storey) is significantly less than the strength of adjacent upper storeys. Plastic hinge formation or brittle failure is concentrated in this storey causing damage and sometimes total collapse of the weak storey. The second is the vertical setbacks. A vertical setback is a horizontal, or near horizontal, offset in the plane of an exterior façade of a structure. Setbacks may be introduced for several reasons. The three most common are zoning requirements that require upper storeys to be set back to admit light and air to adjoining sites, program requirements that require smaller storeys at the upper levels, and stylistic requirements relating to building form.

Along the horizontal direction:

(1) The dimension of the local projection should not be excessively large;

(2) The mass of the lateral-force-resisting members is essentially symmetrically distributed in the plan configuration;

(3) The lateral-force-resisting members should be orthogonal or nearly orthogonal.

For buildings with complex configurations, seismic joints can separate the building structure into simple shapes. Two things should be considered when using a seismic joint. First is that the seismic joint changes the natural period of the structure, which may result in resonance if the changed natural period is similar to the natural period of the ground. Second, the seismic joint must have sufficient width in the vertical; otherwise, the two parts separated by the seismic joint can collide with each other. Nevertheless, a seismic joint with a large width may cause architectural challenges. Occasionally, a seismic joint cannot be used due to structural limitations or architectural functions. In these cases, the structural engineer should perform specific analysis and take appropriate seismic measures by considering the potential stress and strain concentration and torsional response.

4. Energy Dissipation of Structural Members

(1) When selecting structural systems, seismic safety and economic factors must be considered. Different materials and different structural systems have various applicable heights under different earthquake intensities. The relevant codes stipulate the maximum applicable height for masonry structures, reinforced concrete frame structures and frame-shear wall structures.

(2) The natural period of a structural system should not be close to the natural period of the ground; otherwise, resonance may occur. During the Mexico City earthquake in 1985, by analyzing the spectrum of strong earthquake records, it was found that the natural period of the ground is 2 s; thus, structures with a natural period near 2 s are severely damaged by the

earthquake.

(3) There should be more than one seismic line of defense. Certain structural systems may lose their resistance ability due to earthquake or gravity loads after certain components fail. To avoid this type of failure, a structural system should include different systems with ductility, and the ductile members should cooperate. For example, in frame-shear wall structures, the first seismic line of defense is the shear wall, which bears all of the seismic action. The second seismic line of defense is the frame, which can bear some of the shear force induced by the earthquake after the wall cracks. Another example is the in-filled wall frame structure, where the in-filled wall is the first seismic line of defense and bears the entire earthquake-induced shear force. The frame not only increases the bearing and deformation capacity of the in-filled wall by providing constraints but also bears some of the shear force after the wall cracks. In single-storey industrial buildings, the column brace is the first seismic line of defense, and it bears most of the seismic action in the longitudinal direction. If the brace loses its strength, the seismic action is transferred to the columns without a brace.

Additionally, a structural system should have as much redundancy as possible to ensure that the system does not become a variable mechanism after certain members fail. Generally, these members would yield at the beginning of the earthquake, assimilating and dissipating large amounts of energy; thus, the important members are protected. For example, one of the frame design principles is a strong-column-and-weak-beam design. During an earthquake, the beams yield before the columns and dissipate energy, which increases the deformation capability of the structure. For coupled shear walls, the coupling beam is designed to yield first to protect the wall-column coupling via the same principle.

(4) The integrity and deformation ability are of great importance for a structural system in an earthquake. A structural system's seismic resistance ability is related to its strength, rigidity and ductility. In fact, a system with a high seismic ability should have enough strength and a high ductility or deformation ability. A system with high rigidity but low ductility is easily damaged in an earthquake. For example, a plain masonry building without structural columns performs poorly in earthquakes. Conversely, a system with high ductility but low lateral resisting strength, such as a moment-resisting frame structure, would exhibit a large deformation in an earthquake, which is not good for seismic resistance. Severe damage or even collapse of the reinforcement concrete moment-resisting frame structure is common. Thus, integrity and deformation ability are key factors to avoid collapse in a major earthquake. The following regulations are provided by the ***Code for Seismic Design of Buildings*** (GB 50011—2010) Version 2016 for different systems.

(a) In multi-storey masonry buildings, a sufficient number of reinforced concrete columns should be set. These columns along with the ring beams form a frame system and largely increase a building's integrity and deformation ability by providing constraints for the walls. In the Tangshan earthquake, several multi-storey masonry buildings with these columns and ring beams were damaged but did not collapse.

(b) In reinforced concrete frame and frame-shear wall structures, the strong-shear-and-weak-bending principle should be applied. In the panel zone of the beam-column joint, a confined stirrup is used to increase the shear-resistance ability and avoid brittle shear failure. Furthermore, it is suggested to set the boundary frame or embedded columns at the end of the shear wall and place most of the distribution reinforcement at the boundary element. Thus, the system can prevent initial cracks through the wall and increase its deformation ability.

(c) In steel structures, to prevent the overall instability and local instability is of great importance in the design process.

(d) A structural system is formed by several components connected to each other. To upgrade the deformation ability, integrity must be considered. In a prefabricated roof, the ring beams can be set to enhance the integrity. The connection joint between the industrial building roof and the top of the column, the top of the wall or the corbel demands sufficient strength and deformation ability (such as a steel plate hinge). The strength of the joint should be larger than that of the components connected by the joint, and the strength of the connector anchors should be larger than that of the connectors so that the connection zones can remain elastic and keep the integrity reliable, even when the components yield.

(5) A concentration of plastic deformation in a soft or weak storey should be avoided. Among all urban habitat structural problems, the failure of soft and weak storeys has been responsible for more deaths and destruction. If all of the storeys are approximately equal in strength and stiffness, the entire building deflection under seismic action is distributed approximately equally to each storey. However, due to several limitations, such as service functions, material modules, strength and detailing, certain parts of the structure would have relatively lower seismic resistance ability. In a major earthquake, these parts would yield first and cause plastic deformation concentrations.

The *Code for Seismic Design of Buildings* (GB 50011—2010) Version 2016 introduces the yielding strength coefficient as follows:

$$\xi_i = \frac{V_{yi}}{V_{ei}}$$

where,

V_{yi}——the actual shear bearing yield capacity, which is determined by the actual reinforcement area, material characteristic strength and axial force of the columns and walls;

V_{ei}——the elastic seismic shear force in the storey under major earthquake conditions (assuming that the structure remains elastic).

Based on the distribution line chart of ξ_i with height, the storey with the lowest ξ_i and the rest are the soft-weak storeys. In major earthquakes, these storeys would yield first and have a large elastoplastic storey drift.

5. Non-Structural Components

Non-structural components, which include architectural non-structural components (non-bearing walls: parapet walls, retaining walls and partition walls; decorative components: awnings, ceilings, glass curtain walls, billboards, etc.) and the construction of electromechanical equipment and its connection to structural bodies (elevators, heating and air conditioning systems, pyrotechnic monitoring and firefighting systems, public antennas, etc.) should be designed for seismic resistance.

(1) It should have a reliable connection and anchorage with the main structure to prevent local damage to non-structural components in the earthquake.

(2) Enclosure walls and partition walls shall take into account the adverse effects of structural earthquake resistance.

4.3 Calculation Methodology of Seismic Action

4.3.1 General Requirements

1. Calculation Principles

The location where an earthquake will occur is random. For a certain structure, the direction of the seismic action is arbitrary, and the lateral-force-resisting members may not be strictly orthogonally distributed. These factors should be considered in the seismic action calculation. Additionally, a structure's stiffness center may not be located at the exact same place as its mass center, causing torsion in the structure's system. Finally, for structures in the epicenter area, the influence of the vertical seismic action cannot be ignored. Thus, in accordance with the ***General Code for Seismic Precaution of Buildings and Municipal Engineering*** (GB 55002—2021), seismic action calculation should satisfy the following general requirements:

(1) In general, seismic calculations should be carried out in two main directions of structures. If there are diagonal lateral-force-resistant members (greater than 15 degrees with the main direction), seismic calculations for those members in the diagonal direction should be carried out.

(2) Torsional effect should be taken into account during seismic calculations for lateral-force-resistant members.

(3) Structures with large spans or long cantilevers in the region of the seismic intensity degree not lower than 8, or high-rise buildings, structures full with water, air holder, gas storage holder, etc, in the region of seismic intensity degree of 9, vertical earthquake effects should be taken into account.

For spatial structures and long linear structures with large plane projection scale, the spatial and temporal changes of seismic ground motion should be considered during the calculation of seismic action.

2. Calculation Method

In the ***Code for Seismic Design of Buildings*** (GB 50011—2010) Version 2016, three methods are provided to calculate the seismic action:

(1) The response spectrum method (RSM), which applies to multi-degree-of-freedom systems;

(2) The base shear method (BSM), which considers a multi-degree-of-freedom system as an equivalent single-degree-of-freedom system;

(3) The time history method (THM), which inputs earthquake waves directly to solve the equations of motion and obtain the structure's seismic response.

Furthermore, the national code regulates the application ranges of the three methods:

(1) For buildings with a height of less than 40 m, the deformation is dominated by shear deformation, and the mass and stiffness are evenly distributed vertically; for buildings that can be approximated to a single-degree-of-freedom system, a simplified method, such as the base shear method, should be used;

(2) For other buildings, the response spectrum method is recommended to calculate the seismic action;

(3) For buildings that are extremely irregular and belong in Category A or meet the requirements in Table 4-4, the time history method is necessary to perform additional calculations for frequent earthquakes. When using 3 groups for acceleration time history curves, the calculation result is recommended to be the larger value determined by the time history method and the response spectrum method. When using 7 groups (or more) for the acceleration time history curves, the calculation result should be the average determined by the time history method and the response spectrum method.

Building height range for the time history method　Table 4-4

Intensity and site class	Range of building height (m)
Intensity Ⅶ(7) intensity Ⅷ(8) with site Classes Ⅰ and Ⅱ	>100
Intensity 8 with site Classes Ⅲ and Ⅳ	>80
Intensity Ⅸ(9)	>60

When the time history method is used in the analysis, several strong earthquake records and artificial acceleration time history curves should be selected based on intensity, design seismic group and site class. The number of strong earthquake records should be at least 2/3 of the total number of time history curves. The average response spectrum of the selected ground motions should be statistically consistent with the design response spectrum. The maximum value for the acceleration time history can be obtained using Table 4-5 (the values in the parentheses are used when the basic design acceleration of the ground motion is 0.15 g and 0.30 g). When using the elastic time history method, the structure's base shear force obtained from each time history curve should be at least 65% of the value from the response spectrum method, and the average value from several time history curves should be at least 80% of the value from the response spectrum method.

Maximum values for the seismic acceleration of ground motion used in the time history method (cm/s^2)   Table 4-5

Seismic action	Intensity VI (6)	Intensity VII (7)	Intensity VIII (8)	Intensity IX (9)
Frequently occurring earthquakes	18	35 (55)	70 (110)	140
Rarely occurring earthquakes	125	220 (310)	400 (510)	620

In order to correctly select the time history curve, certain requirements for the three characteristics of ground motion need to be satisfied. In particular, the frequency spectrum characteristic, peak ground acceleration and duration should meet the specifications.

The frequency spectrum characteristic can be determined based on the seismic influence coefficient curve, site class and earthquake group.

The peak ground acceleration can be adopted using Table 4-5. If the structure requires earthquake waves in two directions (two horizontal directions) or three directions (two horizontal directions and one vertical direction), such as a three-dimensional spatial model, the maximum value of acceleration in each direction should be adjusted to the ratio of 1 (horizontal 1) : 0.85 (horizontal 2) : 0.65 (vertical). For the actual earthquake record, an earthquake wave in three directions could be three components in a single record or from different records; however, all records must be statistically consistent with the design response spectrum. For an artificial earthquake wave, the requirements mentioned above should also be satisfied.

In general, the duration of the input acceleration time history curve is 5—10 times the structural period for both real earthquake records and artificial waves.

3. Calculation of the Representative Value of Gravity Loads

When an earthquake occurs, the variable load usually does not reach its standard value. Thus, the representative value of the gravity load should be taken as the sum of the standard value of the weight of a particular structure and its members plus the combined values of the variable loads on the structure. The combination coefficients for the different variable loads can be obtained from Table 4-6. When the hanging weight of a crane with hard hooks is relatively large, the combination coefficient should be used based on the actual conditions.

Combination coefficient values   Table 4-6

Type of variable load		Combination coefficient
Snow load		0.5
Dust load on the roof		0.5
Live load on the roof		Not considered
Live load on the floor, calculated according to the actual state		1.0
Live load on the floor, calculated according to equivalent uniform load	Library, archives	0.8
	Other civil buildings	0.5

Continued Table

Type of variable load		Combination coefficient
Gravity for hanging object of crane	Cranes with hard hooks	0.3
	Cranes with flexible hooks	Not considered

4. Determination of the Seismic Influence Coefficient

The seismic influence coefficient of a building structure can be determined based on the intensity, site class, seismic group, natural period and damping ratio of the structure. The maximum value of the horizontal seismic influence coefficient can be obtained from Table 4-7; the characteristic period can be obtained from Table 4-8 based on the site class and seismic group. When calculating the values for a rarely occurring earthquake, the characteristic period should be increased by 0.05 s.

For structures with a natural period greater than 6.0 s, their seismic influence coefficient should be studied in detail.

Maximum value of the horizontal seismic influence coefficient α_{max} Table 4-7

Earthquake influence	Intensity Ⅵ(6)	Intensity Ⅶ(7)	Intensity Ⅷ(8)	Intensity Ⅸ(9)
Frequently occurring earthquakes	0.04	0.08 (0.12)	0.16 (0.24)	0.32
Rarely occurring earthquakes	0.28	0.50 (0.72)	0.90 (1.20)	1.40

Note: The values in parentheses are used when the design basic acceleration of the ground motion is 0.15 g and 0.30 g.

Characteristic period (s)   Table 4-8

Design group	Site class				
	I_0	I_1	Ⅱ	Ⅲ	Ⅳ
First group	0.20	0.25	0.35	0.45	0.65
Second group	0.25	0.30	0.40	0.55	0.75
Third group	0.30	0.35	0.45	0.65	0.90

The damping adjustment and parameter formations on the building seismic influence coefficient curve (Fig. 4-3) should conform to the following requirements:

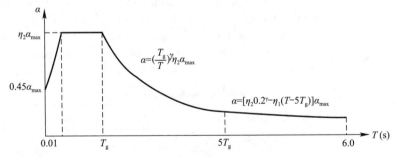

Fig. 4-3 Seismic influence coefficient curve

(1) Except for specific regulations, the damping ratio of a structure should be set to 0.05, the damping adjusting coefficient of the seismic influence coefficient curve should be set to 1.0, and the coefficient of shape should conform to the following provisions:

(a) Linearly increasing section, where the period T is less than 0.1 s;

(b) Horizontal section, where the period ranges from 0.1 s to the characteristic period.

The seismic influence coefficient equals the maximum value α_{max};

(c) Curvilinear decreasing section, where the period ranges from the characteristic period to 5 times the characteristic period. The power index γ should be set to 0.9;

(d) Linearly decreasing section, where the period ranges from 5 times the characteristic period to 6.0 s. The adjusting factor of the slope η_1 should be set to 0.02.

(2) When the damping ratio of a structure is not equal to 0.05 based on the relevant provisions, the damping adjustment and parameter formation on the seismic influence coefficient curve should comply with the following requirements:

(a) The power index of the curvilinear decreasing section can be determined based on the following:

$$\gamma = 0.9 + \frac{0.05 - \zeta}{0.3 + 6\zeta}$$

where,

γ——the power index of the curvilinear decreasing section;

ζ——the damping ratio.

(b) The adjusting factor of the slope for the linearly decreasing section can be determined from the following:

$$\eta_1 = 0.02 + \frac{0.05 - \zeta}{4 + 32\zeta}$$

where,

η_1——the adjusting factor of the slope for the linearly decreasing section; if less than 0, it should equal 0.

(c) The damping adjustment factor can be determined based on the following equation:

$$\eta_2 = 1 + \frac{0.05 - \zeta}{0.08 + 1.6\zeta}$$

where,

η_2——the damping adjustment factor; if smaller than 0.55, it should equal 0.55.

4.3.2 Calculation of the Horizontal Seismic Action

In Chapter 3, Sections 3.8.1 and 3.8.2 have already presented the calculation methods for the response spectrum method and the base shear method, respectively. In the actual design process, the seismic action should be calculated based on the national code, and the method should be selected based on their applicable conditions.

1. Base Shear Method

The base shear method can be used when the following four requirements are met:

(1) The building height is less than 40 m;

(2) The mass and stiffness are evenly distributed along with the building height;

(3) The shear deformation is predominant under earthquakes;

(4) The torsional effect under earthquakes could be neglected.

When the base shear method is used, only one degree of freedom may be considered for each storey. The standard value of the horizontal seismic action for the structure can be determined as follows (Fig. 4-4):

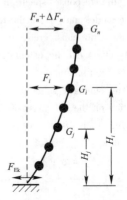

Fig. 4-4 Calculation diagram for the horizontal seismic action

$$F_{Ek} = \alpha_1 G_{eq} \quad (4\text{-}3)$$

$$F_i = \frac{G_i H_i}{\sum_{j=1}^{n} G_j H_j} F_{Ek}(1 - \delta_n)$$

$$(i = 1, 2, \cdots, n) \quad (4\text{-}4)$$

$$\Delta F_n = \delta_n F_{Ek} \quad (4\text{-}5)$$

where,

F_{Ek}——the standard value of the total horizontal seismic action of the structure;

α_1——the horizontal seismic influence

coefficient corresponding to the fundamental period of the structure; for multi-storey masonry buildings and masonry buildings with bottom frames, the maximum value of the horizontal seismic influence coefficient should be used;

G_{eq} ——the equivalent total gravity load of a structure; when the structure is modeled as a single mass system, the representative value of the total gravity load should be used; when the structure is modeled as a multi-mass system, 85% of the representative value of the total gravity load should be used;

F_i ——the standard value of the horizontal seismic action applied on the i-th mass;

G_i, G_j ——the representative value of the gravity load concentrated at the i-th and j-th mass, respectively;

H_i, H_j ——the calculated height of the i-th and j-th mass from the base of the building, respectively;

δ_n ——the additional seismic action factors at the top of the building; for multi-storey reinforced concrete buildings and steel buildings, the value can be obtained using Table 4-9;

ΔF_n ——the additional horizontal seismic action applied at the top of the building.

Additional seismic action factors at the top of the building(s)   Table 4-9

T_g	$T_1 > 1.4 T_g$	$T_1 \leqslant 1.4 T_g$
$T_g \leqslant 0.35$	$0.08 T_1 + 0.07$	
$0.35 < T_g \leqslant 0.55$	$0.08 T_1 + 0.01$	0
$T_g > 0.55$	$0.08 T_1 - 0.02$	

2. Response Spectrum Method

When the response spectrum method is used for the model analysis, if the torsion effect of a structure is not considered, the seismic action and its effect can be calculated based on the following requirements:

(1) The standard value of the horizontal seismic action on the i-th mass of the structure corresponding to the j-th mode can be determined as follows:

$$F_{ji} = \alpha_j \gamma_j X_{ji} G_i$$
$$(i = 1, 2, \cdots n, j = 1, 2, \cdots m) \quad (4\text{-}6)$$

$$\gamma_j = \sum_{i=1}^{n} X_{ji} G_i \Big/ \sum_{i=1}^{n} X_{ji}^2 G_i \quad (4\text{-}7)$$

where,

F_{ji} ——the standard value of the horizontal seismic action of the i-th mass corresponding to the j-th mode;

α_j ——the seismic influence coefficient corresponding to the natural period of the j-th mode of the structure;

X_{ji} ——the horizontal relative displacement of the center of the i-th floor for the j-th mode;

γ_j ——the mode participation factor of the j-th mode.

(2) The total effect of the horizontal seismic action (bending moment, shear, axial force or deformation) when the period ratio between the adjacent modes is less than 0.85 can be determined using the following equation:

$$S_{Ek} = \sqrt{\sum S_j^2} \quad (4\text{-}8)$$

where,

S_{Ek} ——the standard value of the horizontal seismic effect;

S_j ——the effect caused by the horizontal seismic action of the j-th mode; generally, only the first two or three modes are used. When the fundamental natural period exceeds 1.5 s or the height-width ratio of the building exceeds 5, the number of modes used should be increased.

To improve the response spectrum method's calculation accuracy for high-rise and flexible structures, the number of modes involved in the analysis should be adequate. Generally, the number should be sufficient to ensure that the mode participation mass achieves 90% of the total mass.

3. Calculation Method Considering the Torsion Effect

In addition to translational vibrations, torsion can also be generated in structures during an earthquake. There are two reasons for this phenomenon: first is that the ground motion has arotational component or there is a phase difference between the different ground points; second is that eccentricity exists in the structure, i. e. , the stiffness center and mass center do not closely coincide. The studies carried out after earthquakes have shown that the torsional effect aggravates the damage to structures even becoming the primary cause of damage in certain cases. However, no efficient method has yet been proposed to quantitatively calculate the torsional effect caused by the rotational component of the ground motion. In the following section, the torsional effect caused by eccentricity is discussed:

(1) Even for planar regular structures, most seismic design codes consider the torsion effect to be caused by accidental eccentricity and the influence of the rotational component. The Chinese *Code for Seismic Design of Buildings* (GB 50011—2010) Version 2016, which is based on foreign practice and engineering projects in China, proposes methods to consider the torsion effect. When there are no coupled torsion calculations performed for regular structures, the seismic effect of the two-side trusses parallel to the direction of motion induced by an earthquake should be multiplied by an amplifying factor. Generally, the short side uses 1.15, whereas the longer side uses 1.05; when the torsion rigidity is relatively small, it is recommended to select a factor of at least 1.3. For angle components, it is recommended that the seismic effect be multiplied by the coefficients of the two directions at the same time.

(2) When the response spectrum considering torsion is applied, three degrees of freedom may be selected for each floor, including two orthogonal horizontal deformations and a rotation around the vertical axis, and the seismic action and its effects can be calculated based on the following equations. Under certain circumstances, simplified methods may also be used to determine the seismic effect.

(a) The standard horizontal seismic action of the i-th floor for the j-th mode of the natural vibration of the structure can be determined as follows:

$$F_{xji} = \alpha_j \gamma_{tj} X_{ji} G_i$$
$$F_{yji} = \alpha_j \gamma_{tj} Y_{ji} G_i$$
$$(i = 1,2,\cdots,n; j = 1,2,\cdots,m) \quad (4\text{-}9)$$
$$F_{\psi ji} = \alpha_j \gamma_{tj} r_i^2 \varphi_{ji} G_i$$

where,

$F_{xji}, F_{yji}, F_{\psi ji}$——the standard seismic action of the i-th floor for the j-th mode of the natural vibration for the structure in the x and y directions and rotation, respectively;

X_{ji}, Y_{ji}——the horizontal relative displacement of the center of the i-th floor for the j-th mode of the natural vibration for the structure in the x and y directions, respectively;

φ_{ji}——the relative rotation angle of the i-th floor for the j-th mode of the natural vibration for the structure;

r_i——the rotation radius of the i-th floor, which is the square root of the rotating moment of inertia around the center of the i-th floor divided by the mass of this floor;

γ_{tj}——the mode participation factor of the j-th mode considering the rotation effect, which can be determined in accordance with the following equation:

When the earthquake is in the x direction, we have:

$$\gamma_{tj} = \sum_{i=1}^{n} X_{ji} G_i / \sum_{i=1}^{n} (X_{ji}^2 + Y_{ji}^2 + \varphi_{ji}^2 r_i^2) G_i$$
$$(4\text{-}10)$$

When the earthquake is in the y direction, we have:

$$\gamma_{tj} = \sum_{i=1}^{n} Y_{ji} G_i / \sum_{i=1}^{n} (X_{ji}^2 + Y_{ji}^2 + \varphi_{ji}^2 r_i^2) G_i$$
$$(4\text{-}11)$$

When the earthquake diagonal with x, we have:

$$\gamma_{tj} = \gamma_{xj} \cos\theta + \gamma_{yj} \sin\theta \quad (4\text{-}12)$$

where,

γ_{xj}, γ_{yj}——the participation factors defined by Eqs. (4-10) and (4-11), respectively;

θ——the angle between the seismic action direction and the x direction.

(b) The torsion effects of the one-direction horizontal seismic action can be determined using the following equations:

$$S_{Ek} = \sqrt{\sum_{j=1}^{m}\sum_{k=1}^{m}\rho_{jk}S_jS_k} \quad (4\text{-}13)$$

$$\rho_{jk} = \frac{8\sqrt{\zeta_j\zeta_k}(\zeta_j + \lambda_T\zeta_k)\lambda_T^{1.5}}{(1-\lambda_T^2)^2 + 4\zeta_j\zeta_k(1+\lambda_T^2)\lambda_T + 4(\zeta_j^2+\zeta_k^2)\lambda_T^2} \quad (4\text{-}14)$$

where,

S_{Ek}——the torsion effect caused by the standard seismic action;

S_j, S_k——the effect caused by the seismic action of the j-th and k-th modes, respectively; the first 9 ~ 15 modes may be selected;

ζ_j, ζ_k——the damping ratio of the j-th and k-th modes, respectively;

ρ_{jk}——the coupling factor of the j-th and k-th modes;

λ_T——the ratio between the natural periods of the k-th and j-th modes.

(c) The torsion effect of the double-direction horizontal seismic action can be determined based on the larger of the following two values:

$$S_{Ek} = \sqrt{S_x^2 + (0.85S_y)^2} \quad (4\text{-}15)$$

or

$$S_{Ek} = \sqrt{S_y^2 + (0.85S_x)^2} \quad (4\text{-}16)$$

where,

S_x, S_y——the torsion effect caused by the horizontal seismic action along the x and y directions based on the Eq.(4-13), respectively.

4. Seismic Action of a Small Room Projecting from the Roof of a Structure

For a building with a small room projecting from the roof, the seismic action of the small room increases dramatically because of the whipping effect due to the mass and stiffness of the small room (elevator motor room, water tanks, parapets and chimneys) being extremely small compared with that of the lower floors. Furthermore, based on damage reports after earthquakes, small rooms on the roof generally have more significant damage. Thus, the base shear method is not applicable for buildings with a small room projecting from the roof. Their horizontal seismic actions can be calculated using the response spectrum method.

However, in practice, there are several buildings with small rooms on their roof. To simplify the analysis process, the Chinese national code regulates that the base shear method can also be used after making slight modifications. In the calculations, the small room may be considered as one point mass, and its seismic effect can be multiplied by an amplifying factor of 3, which should only be applied to the members of the roof and not the lower part of the structure. However, when designing members connecting to the projecting part, the increased effect should be considered.

5. Regulation of the Minimum Floor Seismic Shear Force

Because the seismic influence coefficient decreases dramatically for long periods, the horizontal seismic action calculated for buildings with a fundamental period larger than 3.5 s may be too small. Additionally, for long-period structures, the primary damage may be caused by the ground motion velocity and displacement; however, the response spectrum method provided in the current **Code for Seimic Design of Buildings** (GB 50011—2010) Version 2016 cannot estimate the influence of these responses. Thus, the limit of the minimum floor seismic shear force is regulated to ensure safety. In the **General Code for Seismic Precaution of Buildings and Municipal Engineering** (GB 55002—2021), the shear factor is provided based on different intensities. The difference in the damping ratio is not considered. The horizontal seismic shear force for each floor of the structure should comply with the requirements of the equation (if not, the calculated value should be adjusted) as follows:

$$V_{Eki} > \lambda \sum_{j=i}^{n} G_j \quad (4\text{-}17)$$

where,

V_{Eki}——the i-th floor shear corresponding to a standard horizontal seismic action;

λ——the minimum seismic shear factor; for weak floors with an irregular vertical structure, these values should be multiplied by an amplifying factor of 1.15; the base values for λ are provided in Table 4-10; for structures with a fundamental period less than 3.5 s or with torsion irregularity, the values of λ should be larger than the benchmark values; for structures with a fundamental period larger than 5.0 s, the values should not be less than 0.75 times of the benchmark values; for structures with a fundamental period between 3.5 and 5.0 s, the values should not less than the benchmark values multiplied by $\dfrac{9.5-T_1}{6}$;

G_j——the representative value of the gravity load for the j-th floor of the structure.

Benchmark value for minimum seismic shear factor   Table 4-10

Intensity	Ⅵ(6)	Ⅶ(7)	Ⅷ(8)	Ⅸ(9)
λ	0.008	0.016(0.024)	0.032(0.048)	0.064

6. Distribution of the Seismic Floor Shear Force

The horizontal seismic shear force at each floor of a given structure should be distributed to the lateral-force-resisting members (such as walls, columns and seismic braces) based on the following principles:

(1) For buildings with rigid diaphragms, such as cast-in-place and monolithic-prefabricated concrete floors and roofs, the distribution should be performed in proportion to the equivalent stiffness of the lateral-force-resisting members.

(2) For buildings with flexible diaphragms, such as wood roofs and wood floors, the distribution should be performed based on the ratio of the gravity load representative value in the areas that are subordinate to the lateral-force-resisting members.

(3) For buildings with semi-rigid diaphragms, such as ordinary prefabricated concrete roofs and floors, the distribution should be applied based on the average value of the above two methods.

(4) The above distribution results may be adjusted in accordance with the relevant provisions in the *Code for Seismic Design of Buildings* (GB 50011—2010) Version 2016 to consider the space interaction of the lateral-force-resistant members, deformation of diaphragms, elastoplastic deformation of the wall and torsion effect.

7. Interaction of the Subsoil and the Structure

In an earthquake, a structure's seismic action is caused by the earthquake wave from the subsoil. Generally, it is assumed that the subsoil is rigid, which is not true in reality. In fact, the seismic action in the upper structure can influence the subsoil through the foundation, causing a local distortion in the subsoil, which results in the entire structure moving and vibrating. This phenomenon is known as the subsoil-structure interaction.

The subsoil-structure interaction changes the subsoil motion and the structural dynamic characteristics in the following aspects:

(1) Changes the content of the frequency spectrum of the subsoil motion; therefore, the area close to the structure's natural frequency is enhanced. Concurrently, the peak ground acceleration is decreased compared with that in the nearby free field.

(2) Prolongs the fundamental period of the structure due to the subsoil flexibility.

(3) Due to the subsoil flexibility, large amounts of energy may be dissipated in the subsoil due to the subsoil hysteresis effect and wave radiation effect. Thus, the structure's vibration decreases. As the subsoil becomes softer, the vibration decreases.

Several studies on the subsoil-structure interaction have indicated that the structure's seismic action should decrease, whereas the deformation and additional internal force caused by the P-Δ effect should increase. The extent of the effect is related to the stiffness of the subsoil and structure. A rigid subsoil has minimal effects on a flexible structure and greater influence on a rigid structure; a soft subsoil has significant effects on a rigid structure but smaller influence on a flexible structure.

Generally, for the seismic computation of a structure, the subsoil-structure interaction can be ignored. However, for tall reinforced concrete buildings in regions of intensities 8 and 9 with site Class III or IV, if the subsoil-structure interaction must be considered, it should comply with the following requirements:

Tall building structures should consist of a box-type or relatively rigid raft foundation or box-pile foundation, and the fundamental period of the structure should be within 1.2 to 5 times the characteristic period of the site.

If a subsoil-structure interaction is considered for these structures, the horizontal seismic shear forces assumed for the rigid base may be reduced in accordance with the following provision, and the storey drift may be calculated based on the reduced storey shear force.

(1) For structures with a height-width ratio less than 3, the reduction factor of the horizontal seismic shear of each floor can be determined as follows:

$$\psi = \left(\frac{T_1}{T_1 + \Delta T}\right)^{0.9} \quad (4\text{-}18)$$

where,

ψ——the seismic shear reduction factor considering the subsoil-structure interaction;

T_1——the fundamental period of the structure (s), which is determined by the assumption of the rigid base;

ΔT——the additional period after considering the subsoil-structure interaction (s), which can be determined from Table 4-11.

Additional period(s)

Table 4-11

Intensity	Site class	
	III	IV
VIII(8)	0.08	0.20
IX(9)	0.10	0.25

(2) For structures with a height-width ratio of at least 3, the seismic shear of the structural bottom may be decreased based on (1) above, the seismic shear of the structural top may not be reduced, and the seismic shear of the middle floors may be reduced based on linearly interpolated values.

(3) The reduced horizontal shear of all floors should comply with Eq.(4-17).

4.3.3 Calculation of the Vertical Seismic Action

Earthquake damage investigations and analysis reports have indicated that horizontal seismic actions cannot explain the damage incurred by certain buildings near the epicenter of high-intensity earthquakes. Based on the measurements and computational analysis performed over the past few years, the ratio of the vertical seismic stress σ_V and gravity load stress σ_G, written as $\lambda_V = \sigma_V / \sigma_G$, increases gradually with height for certain high-rise buildings and structures. For high-rise buildings in intensity 8 zones, the value of λ_V in the upper extent may exceed 1; for chimneys and high structures in intensity 9 zones, the value of λ_V can also achieve or even exceed 1. In these cases, tension stress may exist in the upper part of the structures. Thus, nearly all seismic codes around the world consider vertical seismic action. China's national code states that the vertical seismic action should be analyzed for long cantilever and large-span structures of intensities 8 and 9 as well as high-rise buildings in intensity 9 zones. For different types of structures, the calculation method for the vertical seismic action varies. For chimneys and similar high-rise structures and buildings, the response spectrum method is used to calculate the standard value of the vertical seismic action. For the flat-panel grid and large-span structures, the static method is used to calculate their vertical seismic action. The calculation methods are defined below.

1. Vertical Seismic Action for High-Rise Buildings

The vertical response spectrums corresponding to 203 actual earthquake reports are divided into several groups based on the ground category, and their average response spectrum is calculated. Statistical analysis indicates that the vertical response spectrum is almost the same as the horizontal response spectrum. Thus, in the **Code for Seismic Design of Buildings** (GB 50011—2010) Version 2016, the horizontal response spectrum is used to determine the vertical seismic action. The difference is that the vertical peak acceleration is 65% of the horizontal peak value. As the distance to the epicenter decreases, the ratio increases. Thus, the ratio of the vertical seismic coefficient to the horizontal seismic coefficient $k_V = k_H$ may be selected as 2/3. Thus, the vertical seismic influence coefficient can be calculated as follows:

$$\alpha_V = k_V \beta_V = \frac{2}{3} k_H \beta_H = \frac{2}{3} \alpha_H \approx 0.65 \alpha_H$$

where,

k_V, k_H——the vertical and horizontal seismic coefficients, respectively;

β_V, β_H——the vertical and horizontal dynamic coefficients, respectively;

α_V, α_H——the vertical and horizontal seismic influence coefficients, respectively.

Based on the previous analysis, the deviation is extremely small if the first mode of the vertical seismic action is assumed to represent the entire seismic action for high-rise buildings and structures. The first mode is similar to a straight line; thus, the seismic action in the i-th mass point can be written as follows:

$$F_{Vi} = \alpha_{V1} \gamma_1 Y_{1i} G_i \quad (4\text{-}19)$$

where,

F_{Vi}——the standard value of the seismic action of the i-th mass point;

α_{V1}——the vertical seismic influence coefficient corresponding to the structure's fundamental period, because the vertical fundamental period is relatively small ($T_{V1} = 0.1 - 0.2$ s), the value can be set to $\alpha_{V1} = \alpha_{Vmax}$;

Y_{1i}——the relative vertical displacement of the i-th mass point in the first mode, because the first mode is a straight line, the value can be set to $Y_{1i} = \eta H_i$, where η is the proportion parameter;

G_i——the representative value of the gravity load of the i-th mass point;

γ_1——the participating factor of the first mode, for which the value can be calculated as follows:

$$\gamma_1 = \frac{\sum_{i=1}^{n} G_i Y_{1i}}{\sum_{i=1}^{n} G_i Y_{1i}^2} = \frac{\sum_{i=1}^{n} G_i H_i}{\eta \sum_{i=1}^{n} G_i H_i^2} \quad (4\text{-}20)$$

The standard value of the total vertical seismic action can be written as follows:

$$F_{EVk} = \sum_{i=1}^{n} F_{Vi} = \alpha_{V1} \gamma_1 \sum_{i=1}^{n} Y_{1i} G_i \quad (4\text{-}21)$$

By substituting Eq. (4-20) into (4-21), $\alpha_{V1} = \alpha_{Vmax}$, $Y_{1i} = \eta H_i$, the equation can be written as follows:

$$F_{EVk} = \alpha_{Vmax} \frac{(\sum_{i=1}^{n} G_i H_i)^2}{\sum_{i=1}^{n} G_i H_i^2} = \alpha_{Vmax} \xi' G$$

Let: $\quad G_{eq} = \xi' G \quad (4\text{-}22)$

Then, we have: $F_{EVk} = \alpha_{Vmax} G_{eq}$

where,

G——the representative value of the total gravity load of the structure, $G = \sum_{i=1}^{n} G_i$;

G_{eq}——the equivalent value of the gravity load of the structure;

ξ'——the equivalent gravity load coefficient, which can be set to 0.75 according to the **Code for Seismic Design of Buildings** (GB 50011—2010) Version 2016.

Now, we can calculate the vertical seismic action of each mass point. Using Eq. (4-21), we have the following:

$$\alpha_{V1} \gamma_1 = \frac{1}{\eta \sum_{i=1}^{n} G_i H_i} F_{EVk} \quad (4\text{-}23)$$

By substituting Eq. (4-23) into (4-19) and $Y_{1i} = \eta H_i$, the equation can be written as follows:

$$F_{Vi} = \frac{G_i H_i}{\sum_{i=1}^{n} G_i H_i} F_{EVk} \quad (4\text{-}24)$$

where,

G_i——the representative value of the gravity load for the i-th mass point;

H_i——the height of the i-th mass point.

For high-rise buildings in intensity 9 zones, the vertical seismic action can be calculated using Eq. (4-24) and then multiplied by an amplification factor of 1.5. This requirement is primarily from the experience of the "9·21" earthquake that occurred in the Taiwan region. Generally, the standard value of the total vertical seismic action is larger than the representative value of the gravity load multiplied by 10% and 20% for intensities 8 and 9 zones, respectively.

2. Spatial-Truss Roof or Large-Span Roof Buildings

For flat steel truss roofs with a span length between 24 and 60 m, standard roof trusses with a span greater than 18 m, and large-span structures, the vertical response spectrum method shows that the internal force ratio of the vertical seismic action to the gravity load is nearly invariant. Thus, for flat lattice truss roofs and trusses with a span greater than 24 m as well as a cantilever and other large-span structures, the standard value of the vertical seismic action can be calculated using the static method as follows:

$$F_{Vi} = \lambda_{EV} G_i \quad (4\text{-}25)$$

where,

λ_{EV}——the vertical seismic action coefficient, for flat lattice trusses, steel trusses and reinforced concrete roofs, the value can be determined from Table 4-12; the values in the parentheses are used when the design basic seismic acceleration is 0.3 g;

G_i—— the representative value of the gravity load for the members.

Vertical seismic action coefficients　　　　Table 4-12

Types of structure	Intensity	Site class		
		I	II	III and IV
Plate truss and steel roof truss	Ⅷ(8)	0.0 (0.10)	0.08 (0.12)	0.10 (0.15)
	Ⅸ(9)	0.15	0.15	0.20
Reinforced concrete truss	Ⅷ(8)	0.10 (0.15)	0.13 (0.19)	0.13 (0.19)
	Ⅸ(9)	0.20	0.25	0.25

3. Long Cantilever Beams and the Other Large-Span Structures

For long cantilevers and other large-span structures, the standard vertical seismic action can be taken as 10% and 20% of the representative gravity load for the structure or member in intensities 8 and 9 zones, respectively. When the design seismic acceleration is 0.30 s, the standard vertical seismic action may be taken as 15% of the representative gravity load.

For large-span space structures, the vertical response spectrum method can be used to determine the seismic action. The vertical seismic influence coefficient should be set to 65% of the value listed in the national code, whereas the characteristic period should be selected from the first design group.

4.4　Seismic Check

In China, the seismic design process should follow the "two-stage and three-level" method. In the first stage, the elastic force should be checked under the basic load combination considering minor earthquake action and other load effects; the elastic deformation should be checked under minor earthquake conditions. In the second stage, the elastoplastic deformation should be checked under major earthquake motions. According to the national code, the seismic check for structures should comply with the following provisions:

1) For buildings of intensity 6 (irregular buildings and high-rise buildings in site Class IV except for raw soil and wood buildings), the seismic check can be neglected;

2) For irregular buildings and high-rise buildings in site Class IV of intensity 6 (reinforced concrete frame higher than 40 m, other civil and industrial reinforced concrete buildings higher than 60 m and high-rise steel buildings) and buildings of intensity 7 or higher, a section seismic check under minor earthquake conditions is required.

4.4.1 Seismic Check for the Load-Bearing Capacity of Structural Members

The combination of the responses with other load effects on the structural members can be determined using the following equation:

$$S = \gamma_G S_{GE} + \gamma_{EH} S_{EHk} + \gamma_{EV} S_{EVk} + \psi_w \gamma_w S_{wk} \quad (4\text{-}26)$$

where,

S——the design value of the combination of the inner forces in the structural members, including the design value of the combination of the bending moment, axial force and shear force;

γ_G——the partial factor of the gravity load, which can be taken as 1.3 under ordinary conditions; when the effects of the gravity load are favorable for the bearing capacity of the member, the value should not be greater than 1.0;

γ_{EH}, γ_{EV}——the partial factors for the horizontal and vertical seismic actions, respectively, which can be determined from Table 4-13;

γ_w——the partial factor for the wind load, which can be assumed to be 1.4;

S_{GE}——the effects of the representative value of the gravity load; the standard value of all of the hanging weight of the crane should be included in the effect;

S_{EHk}——the effects of the standard value of the seismic action in the horizontal direction, which can be multiplied by the relevant amplifying factor or adjustment factor;

S_{EVk}——the effects of the standard value of the seismic action in the vertical direction, which can be multiplied by the relevant amplifying factor or adjustment factor;

S_{wk}——the effects of the standard value of the wind load;

ψ_w——the factor for the combined value of the wind load, which can be taken as 0.0 for ordinary structures and 0.2 for tall structures, where the wind load is the control load.

Seismic action partial factors   Table 4-13

Seismic action	γ_{EH}	γ_{EV}
Horizontal seismic action only	1.4	0.0
Vertical seismic action only	0.0	1.4
Horizontal and vertical seismic action (horizontally dominant)	1.4	0.5
Horizontal and vertical seismic action (vertically dominant)	0.5	1.4

The seismic resistance of the cross-section of the structural members can be checked based on the following:

$$S \leq \frac{R}{\gamma_{RE}} \quad (4\text{-}27)$$

where,

γ_{RE}——the seismic adjusting factor of the load-bearing capacity of the structural members, which can be determined from Table 4-14 except for specific requirements; when only the vertical seismic action is considered, the value can be set to 1.0;

R—— the design value of the load-bearing capacity of the structural members, which can be calculated based on the current codes.

Seismic adjusting factor of the load-bearing capacity   Table 4-14

Material	Structural elements	Force condition	γ_{RE}
Steel	Column, beam, brace, connection, bolting and welding	Strength	0.75
	Column, brace	Stability	0.80
Masonry	Confined masonry wall	Shear	0.90
	Masonry wall and reinforced masonry wall	Shear	1.00
Concrete	Beam	Bending	0.75
	Columns with axial force ratio<0.15	Eccentric compression	0.75
	Columns with axial force ratio≥0.15	Eccentric compression	0.80
	Shear wall	Eccentric compression	0.85
	All types of members	Shear, eccentric tension	0.85
Wood	Tension, bending and shear members	Tension, bending and shear	0.90
	Axial compression and compression-bending members	Axial compression and compression bending	0.90
	Timber seismic resistant wall	Strength	0.80
	Connection	Strength	0.85

4.4.2 Seismic Check for Deformation

1. Elastic Deformation Check under Minor Earthquakes

A structure can remain elastic under a minor earthquake in most cases; however, damage may also occur in non-structural components (such as enclosure walls, partition walls and decorations) due to the large elastic deformation. Based on practical research and engineering experiences around the world, the seismic codes in several countries use the storey drift as an indicator to estimate the structure's deformation ability and its capacity to satisfy building functions. In Chinese code, it is provisioned that the maximum elastic storey drift for all types of structures should comply with the following requirement:

$$\Delta u_e \leqslant [\theta_e] h \qquad (4\text{-}28)$$

where,

Δu_e——the elastic storey drift caused by standard minor earthquake action; the storey drift can be computed as the largest difference in the horizontal displacements along any of the edges of the structure at the top and bottom of the storey under consideration except for structures that are dominated by flexural deformation; furthermore, if the torsion effect should be considered, all of the practical factors should be 1.0, and the elastic stiffness can be used for reinforced concrete members;

$[\theta_e]$——the limit of the elastic storey drift rotation can be obtained from Table 4-15;

h——the height of the calculated storey.

2. Elastoplastic Deformation Check under Major Earthquake Conditions

Generally, the peak ground acceleration of a major earthquake is 4—6 times that of a minor earthquake. Thus, most structures reach the elastoplastic stage under a major earthquake even though they remain elastic under minor earthquake conditions.

Limiting value for elastic storey drift   Table 4-15

Types of structure	$[\theta_e]$
Reinforced concrete frame structures	1/550
Dual system	1/800
Reinforced concrete shear wall and tube-in-tube structures	1/1000
Reinforced concrete frame-supporting storeys of structures	1/1000
Multi-storey or high-rise steel structures	1/250

For a structure that is in the elastoplastic stage, there is no more load-bearing capacity. To resist continuous seismic action, the structure needs to have high ductility to dissipate the earthquake input energy through plastic deformation. A structure with low ductility would collapse under a major earthquake. To ensure the structure's safety and that it does not collapse, checking the elastoplastic deformation is required for a major earthquake. The limiting value is provisioned in the code.

1) Scope of the check

An elastoplastic deformation check should be performed for the weak storeys (or location) of the structure under a major earthquake and should comply with the following requirements:

(1) Bent-frames of single-storey reinforced concrete factory buildings with high columns and a large span of intensity 8 with site Classes III and IV or intensity 9;

(2) Reinforced concrete frames and bent-frame structures of intensities 7—9 when the yield strength coefficient of the storey is less than 0.5;

(3) Steel structures with a height greater than 150 m;

(4) Reinforced concrete structures and steel structures that are assigned to Category A or structures assigned to Category B of intensity 9;

(5) Structures using the seismic-isolation and energy dissipation techniques;

An elastoplastic deformation check should be performed for the following structures:

(1) Vertically irregular tall building structures with heights that satisfy Table 4-2;

(2) Reinforced concrete and steel structures assigned to Category B of intensity 7 with site Classes III and IV or intensity 8;

(3) Slab-column-wall structures and masonry buildings with a bottom frame;

(4) Other tall steel structures with heights less than 150 m;

(5) Irregular underground building structures and spaces.

2) Check method

Calculating the elastoplastic deformation under strong seismic motions is an extremely complex task. Thus far, no simple and efficient calculation methods that can be used in actual engineering projects have been established. Overall, there are three deformation estimation methods: (1) calculated by assuming a perfectly elastic body; (2) calculated by the elastic deformation in the seismic action multiplied by an amplification factor; and (3) calculated by specialized programs, such as the time history method. The second method is most commonly used in practice. The 2016 version of the ***Code for Seismic Design of Buildings*** (GB 50011—2010) specifies the content of the seismic check based on the 1989 version.

The elastoplastic deformation of a weak storey (or location) of a structure under a major earthquake can be determined by the following methods:

(1) For reinforced concrete frames and bent-frame structures that do not exceed 12 storeys in height and have no abrupt changes in the storey stiffness and single-storey factory buildings with reinforced concrete columns, the sim-

plified method can be used;

(2) For other structures, the static elastoplastic analysis method and the elastoplastic time history method may be used;

(3) Regular structures may use the bending shear model or the planar line-member system model, and irregular structures should use the spacious structure model.

In the simplified method, weak storeys (or locations) must be initially determined. The structure's weak storey is the storey that yields first and has large elastoplastic deformations under strong earthquake effects. For multi-storey and high-rise buildings, the national code uses a yield strength coefficient to determine the weak storeys. The yield strength coefficient can be determined using the following equation:

$$\xi_y(i) = \frac{V_y(i)}{V_e(i)} \quad (4\text{-}29)$$

where,

$\xi_y(i)$——the yield strength coefficient of the i-th storey;

$V_y(i)$——the shear bearing capacity of the i-th storey, which can be determined based on the actual reinforcement area and the material standard strength of the members;

$V_e(i)$——the elastic seismic shear force of the i-th storey, which can be calculated using the standard value of the seismic action under major earthquake conditions.

For bent-frame structures, where bending is predominant, the yield strength coefficient refers to the ratio of the bending capacity to the elastic seismic moment. The bending capacity can be determined based on the actual reinforcement area, the standard value of the material strength and axial force; the elastic seismic moment can be determined based on the standard value of the major earthquake effect.

The weak storey (or location) can be identified as follows:

(1) For a structure with a uniform distribution in the storey yield strength coefficient in the vertical, the first storey of the building is the weak storey.

(2) For a structure with a non-uniform distribution in the storey yield strength coefficient in the vertical, the storey (or location) with the minimum or relatively small storey yield strength coefficient should be denoted as the weak storey. However, there should be at most two or three storeys (or locations) that are identified as weak storeys.

(3) For single-storey factory buildings, the weak location is at the upper portion of the columns.

The so-called "uniform distribution in the storey yield strength coefficient in the vertical" indicates that the yield strength coefficient of the weak storey (location) does not exceed 80% of the average value of the coefficients of the neighboring storeys (locations), which can be represented as follows:

$$\xi_y(i) > 0.8[\xi_y(i-1) + \xi_y(i+1)]/2$$
(Standard storey) $\quad (4\text{-}30)$

$$\xi_y(n) > 0.8\xi_y(n-1)$$
(Top storey) $\quad (4\text{-}31)$

$$\xi_y(1) > 0.8\xi_y(2)$$
(Bottom storey) $\quad (4\text{-}32)$

Otherwise, the yield strength coefficient is non-uniformly distributed.

The elastoplastic storey drift can be calculated as follows:

$$\Delta u_P = \eta_P \Delta u_e \quad (4\text{-}33)$$

or

$$\Delta u_P = \mu \Delta u_y = \frac{\eta_P}{\xi_y} \Delta u_y \quad (4\text{-}34)$$

where,

Δu_P——the elastoplastic storey drift;

Δu_y——the yield storey drift;

μ——the storey ductility factor;

Δu_e——the elastic storey drift under major earthquake conditions;

η_P——the amplification factor for the elastoplastic storey drift; when the yield strength coefficient of the weak storey (location) does not exceed 80% of the average value of the coefficients of the neighboring storeys (location), the value can be obtained from Table 4-14; when the yield strength coefficient does not exceed 50% of the above-mentioned average value, the value can be determined by the corresponding values in Table 4-16 after multiplying by 1.5; when the yield strength coefficient is between the above two cases, the value can be determined by interpolation;

ξ_y——the yield strength coefficient.

Amplification factor for the elastoplastic storey drift   Table 4-16

Types of structure	Total storeys or locations	η_P		
		0.5	0.4	0.3
Multi-storey frame structure with uniform elevation	2—4	1.30	1.40	1.60
	5—7	1.50	1.65	1.80
	8—12	1.80	2.00	2.20
Single-storey factory buildings	Upper portion of column	1.30	1.60	2.00

The elastoplastic storey drift in the weak storeys (locations) of a structure should comply with the following requirement:

$$\Delta u_P \leq [\theta_P] h \qquad (4\text{-}35)$$

where,

h——the height of the weak storey (location) or the height of the upper portion of the column in a single-storey factory building;

$[\theta_P]$——the limiting value of the elastoplastic storey drift rotation, which can be obtained from Table 4-17. For reinforced concrete frame structures, the values may increase. For example, the value may increase by 10% if the axial force ratio is less than 0.40, and the value may increase 20% if the standard stirrup value along the full height of the column is 30% greater than the minimum stirrup value provisioned in the national code; however, the total increase should not be greater than 25%.

Limiting values of the elastoplastic storey drift   Table 4-17

Types of structures	$[\theta_P]$
Reinforced concrete frame structures	1/50
Dual system	1/100
Frame-supported system	1/100
Reinforced concrete shear wall and tube-in-tube buildings	1/120
Multi-storey or high-rise steel structures	1/50

Chapter 5
Seismic Design of RC Structures

5.1 Introduction

Reinforced concrete (RC) structures have been widely used in multi-storey and high-rise buildings in China, especially in the seismic risk zones. Depending on building height and seismic intensity, different RC structural systems (e.g., frame structures, frame-shear wall structures, shear wall structures, tube-in-tube structures, bundled tube structures, etc.) are appropriate in different situations. Furthermore, the frame structures with specially shaped columns have become popular and accepted among occupants due to their advantages for the occupation of interior space, i.e., the absence of protruding interior columns.

The frame structures can achieve a more flexible division of space, which is beneficial for the plane configuration. Thus, they are commonly used in single-storey and multi-storey RC buildings, such as residences, public infrastructures and industrial factories. However, the taller the building is, the larger the lateral deformation due to earthquake and wind. Enlarging the size of structural members is the most straightforward solution, although it is not economical. In practice, pure frame structures are often replaced by shear walls or frame-shear walls for buildings higher than 10 storeys.

When the architectural function is regular, the resistance to lateral forces may be assigned entirely to the RC or masonry-structural walls (shear walls). The gravity load effects on such walls are seldom significant, and they do not control the design. The contribution of other elements within such a building to the lateral force resistance, if any, is often neglected. The main advantages of the shear wall structure are (1) high integrity, (2) high lateral stiffness and (3) high net storey height.

The frame-shear wall structure, also known as the dual system, combines the advantages of the above two structural systems: the shear wall is the major resistance to lateral loads, whereas the frame is the major resistance to vertical loads. Ductile frames interacting with walls are capable of providing significant amounts of energy dissipation, particularly in the upper storeys of a building. In addition, inter-storey drifts are more evenly distributed over the elevation of the structure due to the large stiffness of walls, and the potential storey mechanisms involving column hinges in frame structures can be avoided. The deformation patterns of the aforementioned structures are shown in Fig. 5-1.

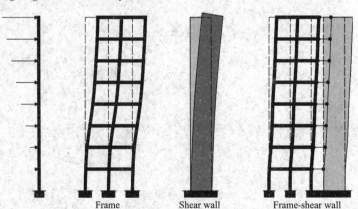

Fig. 5-1 Deformation pattern of a frame, a wall and a frame-shear wall subjected to lateral seismic loads

5.2 Earthquake Damage and Analysis

An on-site investigation and damage analysis are key capabilities of structural engineers. As stated in a Chinese idiom, knowledge comes from practice and returns to practice. Serious losses caused by earthquakes always trigger engineers' introspection, and lessons learned from past earthquake disasters propel the modifications of seismic design codes.

5.2.1 Damage to Frame Members

The majority of frame damage is limited to the joints. In general, (1) columns are more damageable than beams; (2) corner columns are more damageable than interior columns; (3) short columns are more damageable than normal columns; (4) bottom storeys are more damageable than upper storeys;

The typical damage patterns of RC frame members are summarized in Table 5-1 (Fig. 5-2).

5.2.2 Damage to Shear Walls

There are two aspects of structural wall damage:

(1) Shear damage to structural walls is primarily located in the coupling beam when its aspect ratio is less than 1.5 (e.g., X-shaped cracks in deep beams);

(2) Horizontal cracks propagate along the top or bottom of the structural walls without boundary elements, especially when the thickness of the wall is small (aspect ratio larger than 2) or the horizontal reinforcement is under-configured.

Summary of the seismic damages for frame members   Table 5-1

Member	Location	Characteristics	Reasons
Column	Top	Horizontal or diagonal cracks	Large and complex stresses at the top
Column	Top	Concrete crushing	Large and complex stresses at the top
Column	Top	Reinforcement yielding (lantern shape)	Large and complex stresses at the top
Column	Bottom	Horizontal cracks	Lack of stirrups; Loose concrete
Column	Bottom	Concrete spalling	Lack of stirrups; Loose concrete
Column	Bottom	Reinforcement yielding	Lack of stirrups; Loose concrete
Column	Other location	Diagonal cracks	Lack of sufficient shear capacity
Beam	Two ends	Vertical cracks	Concrete cracks under repeated cycle loads
Beam	Two ends	Diagonal cracks	Lack of stirrups
Joint	—	Diagonal cracks and concrete crushing	Large and complex stresses; Loose concrete due to limited space
Joint	—	Stirrup yielding	Large and complex stresses; Loose concrete due to limited space

Fig. 5-2　Damage to frame members

5.2.3　Damage to In-Filled Walls

Shear and flexural failure can occur in in-filled walls. In particular, in a frame structure system, in-filled walls are more severely damaged in the lower storeys than the upper storeys because of the comparatively larger deformation and shear. Conversely, in a dual system, in-filled walls in the upper storeys are subjected to more lateral deformation, resulting in more damage than those in lower storeys.

5.2.4　Other Damages

(1) High-rise buildings on soft soil are prone to severe damage even under a moderate earthquake, as resonance can amplify the seismic response and cause unexpected damage when the natural period of the structure is close to the natural period of the soil.

(2) Frame structures on soft soil or a liquefiable soil layer are more likely to have a global inclination and even collapse in earthquakes because the uneven settlement is unfavorable for load transmission through the entire structural system.

(3) Adjacent structural members separated by a seismic joint have a large relative displacement due to the different vibrations of the two substructures. Thus, collision and damage can occur when the width of the gap is inadequate.

(4) A drastic change in the vertical stiffness causes deformation concentrated in the weak storey and induces global collapse.

5.3　Structural Systems and Seismic Grading

5.3.1　Selection of Structural System

The selection of a structural system is extremely important during the initial design of a building. The building and its structure should have a uniform and continuous distribution of mass, stiffness, strength and ductility. The desirable aspects of the building form are simplicity, regularity and symmetry both in plan and elevation. Also considering the seismic intensity, site category, earthquake-resistant properties of superstructure systems, service functions and economic factors, the height limits for RC buildings are specified in Chinese code and are shown in Table 5-2. The height of the superstructure is defined as the height from the outdoor ground level to the apex of the main roof (not including protruding parts over the roof). The limits in the table are not actual limitations, and the relatively appropriate heights are based on both seismic safety and economic factors. It should be understood that a structure is a unit of material, height and system with consideration of safety and economy.

Appropriate maximum height for RC buildings (m) Table 5-2

Structure types	Seismic intensity				
	VI(6)	VII(7)	VIII(8)(0.2 g)	VIII(8)(0.3 g)	IX(9)
Frame structures	60	50	40	35	24
Frame-structural wall structures	130	120	100	80	50
Structural wall structures	140	120	100	80	60
Partial frame-support structural wall structures	120	100	80	50	Forbidden
Frame-core tube structures	150	130	100	90	70
Tube-in-tube structures	180	150	120	100	80
Slab-column-wall structures	80	70	55	40	Forbidden

Furthermore, other issues should be taken into account:

(1) When the building height is given, the appropriate structure type can be selected based on the table above.

(2) The natural period of a building should not coincide with the period of the ground to avoid resonance.

(3) Proper selection of the foundation system is essential. The foundation should be sufficiently deep, and a raft foundation via a basement may be an alternative solution when building on potentially liquefiable layers. The removal of soil to form the basement tends to reduce the overall inertia of the soil, and the superstructure that must be erected at the foundation level. Pile, raft and basement foundations are the preferred forms when buildings are erected on soft soil layers.

(4) When the building height exceeds 180 m, it is recommended to use the core tube-outrigger structure, mega column-core tube-outrigger structure or mega bracing-core tube-outrigger structure.

5.3.2 Structural Configuration

(1) The principles of the plane assignment are simplicity, regularity and symmetry. Irregular configurations in Table 5-3 are not recommended.

Plane structural irregularities Table 5-3

Type of irregularity	Definitions
Torsion irregularity	The maximum elastic floor displacement or inter-storey drift is more than 1.2 times the corresponding average of two ends of the floor
Concave irregularity	The projection beyond a reentrant corner is greater than 30% of the total plan dimension in the given direction
Diaphragm discontinuity	The dimensions and stiffness of the diaphragm change abruptly, including those having the effective width of diaphragm less than 50% of the typical width, the cutout or open area greater than 30% of the gross enclosed floor area, or staggered floor

(2) The configuration and elevation of a building should be regular, and the lateral stiffness should be changed as evenly as possible. Furthermore, the cross-sectional dimensions and

the material strength of the vertical lateral-force-resisting members should be gradually reduced along with the entire structure from the lower part to the upper part to avoid soft storeys and weak storeys in the structure. Additionally, Table 5-4 lists a few elevation irregularities that should be avoided.

Vertical structural irregularities Table 5-4

Type of irregularity	Definitions
Stiffness irregularity	The lateral stiffness is less than 70% of that in the upper storey or less than 80% of the average stiffness of the upper three storeys; the horizontal dimension is less than 75% of that of the lower storey except for the top storey of the building
Discontinuity in vertical lateral-force-resisting members	The internal forces of the vertical lateral-force-resisting members (columns, structural walls and braces) are transferred to the lower members using the horizontal transmission member (girders or trusses)
Discontinuity in bearing capacity	The storey lateral shear capacity is less than 80% of the adjacent storey above. The storey lateral shear capacity is the total capacity of the anti-lateral-force members sharing the storey for the direction under consideration

(3) To avoid torsional damages, it is recommended that the frames and structural walls be arranged in both the principal directions and that the central axes of the structural members overlap and that the maximum offset of the central axes between beams and columns is 25% of the column width.

(4) Integrity and connection with the structural walls are extremely important for the assembly of precast diaphragms. In general, the depth of a RC layer over the precast diaphragms should be at least 50 mm.

(5) To ensure adequate structural redundancy, a single-span frame is prohibited for classes 1, 2 and 3 buildings higher than 24 m; furthermore, it is also not recommended for Class 3 buildings that are shorter than 24 m because a single-span frame is liable to collapse when plastic hinges develop on the columns.

(6) In order to allow the floor or roof diaphragm to effectively transmit lateral shear to structural walls for frame-wall structures and slab-column-wall structures, the aspect ratio of the floor slab between the adjacent walls should, if possible, be limited to the values suggested in the ***Code for Seismic Design of Buildings*** (GB 50011—2010) Version 2016 provided in Table 5-5. Otherwise, the diaphragm deformation of the floors or roofs should be considered.

Aspect ratios of the diaphragm between two adjacent structural walls Table 5-5

Slab types		Seismic intensity			
		VI(6)	VII(7)	VIII(8)	IX(9)
Frame-wall structure	Cast-in-place or composite slab	4	4	3	2
	Assembling precast slab	3	3	2	N/R
Cast-in-place slab in the slab-column-wall structure		3	3	2	—
Cast-in-place slab in the frame-supported storey		2.5	2.5	2	—

(7) The foundations under the frame columns should be connected to the foundation beams or coupling beams in both principal directions.

(8) For a building with an integral tower and podium, the natural foundation should be checked as discussed in Chapter 2, and the zero-stress zone should be prohibited.

Furthermore, some commonsense judgments from structural engineers can be given as follows:

(1) A low center of gravity and lightweight structure are advantageous to minimize the total seismic force. Thus, lightweight materials are commonly used in partition and enclosure walls, and the base room or bottom storey is used for heavy building facilities.

(2) A cast-in-place staircase is highly recommended. For frame structures, the arrangement of staircases should meet the regularity requirement of the structural plane. Cast-in-place staircases should be checked for seismic forces and should include appropriate structural detailing. The in-filled walls in a staircase should be firmly connected to the columns.

(3) The elevations of the beams in a storey should coincide. However, if there are distinct storeys or short columns, the design should be carefully checked with the codes.

(4) When the top slab of the basement is used as the fixing location of the structural system, the opening of large holes on this diaphragm should be avoided. Additionally, the floor should use the cast-in-situ structure and comply with the following requirements:

(a) The thickness of the slab should be at least 180 mm and the concrete strength grade should be at least C30. Double-layer and bidirectional reinforcements should be used. Moreover, the ratio of reinforcement should be at least 0.25%.

(b) The lateral stiffness of the basement structure should be at least 2 times that of the first storey of the building.

(c) The area of the longitudinal bars on each side of the basement column cross-section should be at least 1.1 times the area of the corresponding column in the first storey of the building. The bending moment design values of the framed columns and the bottom section of the walls in the first storey should comply with the provisions in Sections 6.2.3, 6.2.6 and 6.2.7 of the *Code for Seismic Design of Buildings* (GB 50011—2010) Version 2016.

(d) The actual bending capacity of the columns at the upper face in the joint at the top of basement columns should be less than the sum of the actual bending capacity of the beam at the left and right faces in that joint and the columns at the lower face in that joint.

5.3.3 Seismic Joints

Tall RC buildings should avoid using a structure with seismic joints. When seismic joints are necessary, they should satisfy the following requirements:

(1) The minimum clear width of the seismic joint should comply with the following:

(a) For frame structures, when the building height is at most 15 m, a joint with a width of 100 mm may be used. When the frame height is more than 15 m, the width should be increased by 20 mm for every 5 m, 4 m, 3 m and 2 m increase in height for intensities 6, 7, 8 and 9, respectively.

(b) For frame-wall structures, this width may be taken as 70% of the values as denoted in Item (a); furthermore, neither should be smaller than 100 mm.

(c) When the structural systems at the two sides of the seismic joint are different, the width should be determined based on the wider structural system and the lower building height.

(2) For intensities 8 and 9, when the total height, stiffness or storey height of the frame structures at the two sides of the seismic joint are significantly different, a retaining wall may be installed at the structural edges of the two sides of the seismic joint. This retaining wall should comply with the following requirements:

(a) Along with the total height of the

structure and orthogonal to the seismic joint.

(b) Retaining walls on each side should consist of at least two segments, symmetrically arranged, and the length of this wall should not exceed half the span of the columns.

(c) The internal force of the frame and the retaining wall should be analyzed based on the two cases with and without retaining walls, and the choice can be made based on the more unfavorable situation.

(d) For the end columns of the retaining wall and the side columns of the frame, the spacing of the hoops/spirals of the columns should be condensed along with the total height of the building.

The seismic joints must be installed over the entire height of the above-ground part of a building (superstructure). Normally, seismic joints can be designed in relation to the thermal and settlement joints.

5.3.4 Seismic Grading

An appropriate seismic grading system can balance safety and economic factors. For multi-storey RC buildings, (1) different structural systems have different seismic performances (e.g., the frame-wall system has a better anti-collapse ability than the pure frame system); (2) different structural members have different importance (e.g., frame members in the pure frame system, where they provide the primary seismic resistance, are much more important than those in the frame-wall system, where they provide secondary seismic resistance); (3) as the height of the building increases, the seismic response increases, which indicates an increased ductility requirement.

The seismic grading terminology for structural systems in the Chinese **General Code for Seismic Intensity Buildings and Municipal Engineering** (GB 55002—2021) is based on (1) the seismic intensity, (2) the seismic fortification category, (3) the structural types and (4) building height, which are divided into four seismic grades (Grade 1 represents the highest requirement). Table 5-6 provides a seismic grading system for seismic check and seismic detailing measures of cast-in-place RC buildings.

Seismic grade of RC buildings   Table 5-6

Structural types		Seismic intensity							
		VI(6)		VII(7)		VIII(8)		IX(9)	
Frame structure	Height(m)	≤24	25—60	≤24	25—50	≤24	25—40	≤24	
	Frame	4	3	3	2	2	1	1	
	Large-span frame	3		2		1		1	
Frame-wall structure	Height(m)	≤60	60—131	≤24	25—60	61—120	≤24	25—60	61—100	≤24	25—50
	Frame	4	3	4	3	2	3	2	1	2	1
	Structural wall	3		3	2		2	1		1	
Structural-wall structure	Height(m)	≤80	81—140	≤24	25—80	81—120	≤24	25—80	>80	≤24	25—60
	Structural wall	4	3	4	3	2	3	2	1	2	1

Continued Table

Structural types			Seismic intensity						
			VI(6)		VII(7)		VIII(8)		IX(9)
Frame-supported-wall structure	Height(m)		≤80	81—120	≤24	25—80	81—100	≤24	25—80
	Structural wall	General	4	3	4	3	2	3	2
		Strengthened	3	2	3	2	1	2	1
	Frame		2		2		1	1	
Frame-tube structure	Height(m)		≤150	151—220	≤130	131—190	≤100	101—170	≤70
	Frame		3	2	2	1		1	1
	Tube		2		2	1	1	S1	S1
Tube-in-tube structure	Height(m)		≤180	181—280	≤150	151—230	≤120	121—170	≤90
	Exterior tube		3	2	2	1		1	1
	Interior tube		2		2	1	1	S1	S1
Slab-column-wall structure	Height(m)		≤35	36—80	≤35	36—70	≤35	36—55	
	Column		3	2	2	2		1	
	Wall		2	2	2	1	2	1	

Additionally, the following aspects should be taken into account:

(1) For site Class I, the grades should be reduced by one unit based on the list in Table 5-6 except for intensity 6. However, the design requirements should not be reduced.

(2) When the building height is close to or equal to the height threshold, the grade should be adjusted appropriately depending on the extent of building irregularities and the site and basement conditions.

(3) The long-span frame in Table 5-6 represents a frame with a span exceeding 18 m.

(4) Frame-tube structures lower than 60 m should comply with the provisions for frame-structural wall structures.

(5) For frame-wall structures subjected to the fundamental mode of seismic actions, when the seismic overturning moment distributed to the frame parts is more than 50% of the total seismic overturning moment of the structure, the seismic grade of the frame parts for such structures should be determined in a manner analogous to that of framed structures.

(6) When the podium is connected to the main building, the seismic grades should be determined with the podium itself and be at least that of the main building. Furthermore, a detailed design of the main building should be appropriately strengthened at the level of the top of the podium and the adjacent upper and lower levels. When the main building and the podium are separated, the seismic grades should be determined based on that of the podium.

(7) When the top slab of the basement is used as the fixing location for the structural system analysis, the seismic grade of the first storey below ground should be the same as the

structural system. In addition, the seismic grade for other storeys below the first basement storey may use Grade 3 or Grade 4 based on the actual conditions. For parts of the basement without a corresponding structural system, the seismic grade should be Grade 3 or Grade 4.

(8) For a building in fortification Categories A and B, the seismic grades can be determined for one intensity level higher. Furthermore, when the building height is over the maximum height mentioned in Table 5-6, an additional seismic analysis should be considered.

5.4 Seismic Design of RC Frames

5.4.1 General Design Scheme

Compared to the non-seismic design process, the dynamic properties and the seismic performance should be carefully considered, e.g., the natural period of a structure should be as far as possible from the natural period of the site. Moreover, seismic cases should be combined with wind and gravity cases in loading compositions.

A general design flow chart is depicted in Fig. 5-3; this flow chart is suitable not only for RC structures, but also suitable for other structures including steel and masonry structures.

5.4.2 Seismic Response Calculation

As discussed in detail in Chapter 3 and Chapter 4, there are three primary computation methods for computing the seismic action of a structure: (1) the base shear method (BSM), (2) the response spectrum method (RSM) and (3) the time history method (THM). For the BSM, the following four prerequisites are repeated: (1) the building height should not exceed 40 m, (2) the mass and stiffness are regularly distributed in the vertical, (3) the shear deformation is the predominant pattern under earthquake motions, and (4) the torsional effect under earthquake motions can be neglected.

Section 5.4.1 states that the natural structural period is an essential factor in seismic design; however, we cannot always obtain the exact periods because of the inevitable gap between the computation model and the actual structure. This gap is extended by errors from modeling simplifications, structural uncertainties, etc. Fortunately, the approximate values of these parameters already satisfy the structural design.

There are two primary approaches for estimating natural periods: computer-based dynamic analysis and empirical engineering formulas.

(1) Computer-based dynamic analysis relies on structural modeling inputs and analysis codes. Generally, the computational accuracy depends on the fidelity of the stiffness and the mass matrices. In order to achieve a balance between accuracy and resource consumption (e.g., time and RAM), different objectives should be considered for all researchers and civil engineers performing simulations and structural analyzes.

(2) The empirical engineering formulas are based on a large number of observations and data from existing buildings or experiments. Here are several useful rules:

(a) Non-industrial frame and frame-wall structures:

$$T_1 = 0.33 + 0.00069 \frac{H^2}{\sqrt[3]{L}} \quad (5\text{-}1)$$

where,

H——the overall structural height excluding the protruding parts over the roof (m);

L——the width of the building along the vibration direction (m).

(b) Multi-storey RC industrial structures:

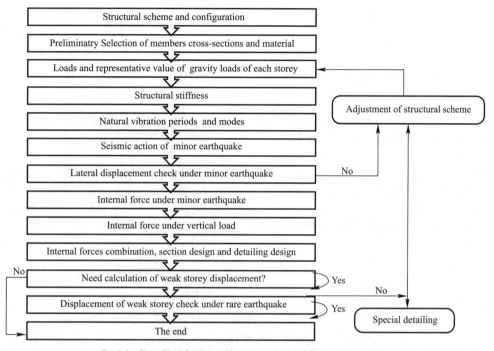

Fig. 5-3 Flow Chart for the multi-storey and tall building structural design

$$T_1 = 1.25 \times \left(0.25 + 0.00013 \frac{H^{2.5}}{\sqrt[3]{L}}\right) \quad (5\text{-}2)$$

where,

H——the equivalent building height calculated using Eq. (5-3) (m);

L——the equivalent width of the building along the vibration direction calculated using Eq. (5-4) (m).

$$H = H_1 + \left(\frac{n}{m}\right) H_2 \quad (5\text{-}3)$$

$$L = L_1 + \left(\frac{n}{m}\right) L_2 \quad (5\text{-}4)$$

The coefficients H_1, H_2, L_1, L_2, n and m are illustrated in Fig. 5-4.

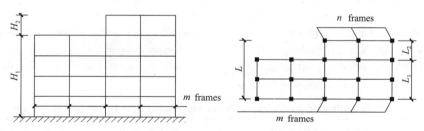

Fig. 5-4 Equivalent height and width for frame structure evaluation

Just like "Big Data" or "Cloud Data", these formulas can be considered reliable for statistical reasons. However, the accuracy is highly influenced by site conditions and structural plans. Thus, empirical engineering formulas are almost always used to verify either the numerical analysis results or the preliminary stage results.

5.4.3 Internal Force under Seismic Action

As introduced in Chapter 4, the seismic action of each storey can be calculated using either the BSM or RSM. Then, the total shear of the storey can be determined using Eq. (5-5) for the BSM and Eq.(5-6) for the RSM:

$$V_i = \sum_{k=i}^{n} F_k + \Delta F_n \tag{5-5}$$

$$V_i = \sqrt{\sum_{j=1}^{m}\left(\sum_{k=i}^{n} F_{jk}\right)^2} \tag{5-6}$$

where,

ΔF_n——the horizontal seismic effects of the top considering the whipping effect;

F_{jk}——the seismic action of the k^{-th} column in the j^{-th} mode.

Based on the shear derived from Eq.(5-5) or Eq.(5-6), the shear of a column in the storey can be determined using the D-value Method as follows:

$$V_{ik} = \frac{D_{ik}}{\sum_{k=1}^{n} D_{ik}} V_i \tag{5-7}$$

where,

V_{ik}, D_{ik}——the shear and stiffness of the k^{-th} column in the i^{-th} floor, respectively.

The lateral stiffness D can be calculated as follows:

$$D = \begin{cases} D_0 = \dfrac{12 i_c}{h^2}, \bar{i} \geqslant 3 \\ \alpha D_0 = \dfrac{12 \alpha i_c}{h^2}, \bar{i} < 3 \end{cases} \tag{5-8}$$

where,

i_b, i_c——the ratios of the flexural stiffness and the effective length of a beam and a column, i.e., Eqs.(5-9) and (5-10), respectively;

α——the joint rotation coefficient, which is influenced by the beam-to-column stiffness, restraint condition of the column ends, etc. (Table 5-7).

Values of $\bar{i}$ and α Table 5-7

Storey	Exterior column	Interior column	α
First storey	$\bar{i} = \dfrac{i_{b5}}{i_c}$	$\bar{i} = \dfrac{i_{b5} + i_{b6}}{i_c}$	$\alpha = \dfrac{0.5 + \bar{i}}{2 + \bar{i}}$
Others	$\bar{i} = \dfrac{i_{b1} + i_{b3}}{2 i_c}$	$\bar{i} = \dfrac{i_{b1} + i_{b2} + i_{b3} + i_{b4}}{2 i_c}$	$\alpha = \dfrac{\bar{i}}{2 + \bar{i}}$

In Table 5-7, i_{b1} through i_{b6} represent the stiffness per length of the beams; i_c is the stiffness per length of the columns. When calculating the moment of inertia of beam I_0, the restraint of the slab is always neglected. However, the slab works with the beam, similar to the flange in a T-shaped beam. To adjust the value, we introduce an amplification coefficient (Table 5-8), which is related to the location and construction method of the slab. However, for precast frame structures, if there are no reliable anchorages between the slab and the beam, the amplification coefficient is equal to 1.0.

Amplification coefficient of the inertia moment for frame beams   Table 5-8

Construction type	Interior frame	Exterior frame
Cast-in-place beam slab	2.0	1.5
Precast composite beam slab	1.5	1.2

The elastic modulus of concrete decreases when a nonlinear elastoplastic deformation occurs, which subsequently indicates a loss in stiffness. Instead of using the initial stiffness, the stiffness reduction coefficient β (Table 5-9) is introduced, which is related to the construction type and the location of the structural members. The adjustment may not be obvious for internal forces, but it is really effective for estimating displacement.

Stiffness reduction coefficient β Table 5-9

Construction type	Frame and structural wall	Coupling beam
Cast-in-place	0.65	0.35
Precast	0.50—0.65	0.25—0.35

The moment profile induced by the shear of each column in a storey can be derived assuming equilibrium as long as the position of the point of contraflexure in the columns is determined. If the stiffness ratio of the beam to the column exceeds 3, the position of contraflexure is at the height of approximately $2h/3$ for the bottom storey and $h/2$ for the other storeys. When the stiffness ratio of the beam to the column is less than 3, the position of the point of contraflexure can be derived based on the D-value method, which can be written as follows:

$$h' = (y_0 + y_1 + y_2 + y_3)h \quad (5-9)$$

where,

y_0——the coefficient determined by the stiffness ratio of the beam to the column, the number of storeys and the storey number (Table 5-10);

y_1——the adjustment coefficient allowing for beams with different stiffnesses above and below the column (Table 5-11);

y_2, y_3——the adjustment coefficients based on the point of contraflexure when the storey height above or below is different from that of the storey of interest (Table 5-12).

When the contraflexure point is determined, the moment of the column can be calculated as follows:

$$M_{kl} = V_{ik}h' \quad (5-10)$$
$$M_{ku} = V_{ik}(h - h') \quad (5-11)$$

where,

M_{kl}, M_{ku}——the design value of both ends of the column;

V_{ik}——the design value of the shear force;

h, h'——the height of the contraflexure point and the floor, respectively.

With the column moments derived from Eqs. (5-10) and (5-11), the total moment of the beams can be derived, and the total moment of the beams is distributed on each beam that connects to the column and is proportional to the beam stiffness. The shear forces acting on each beam induced by the seismic action can be calculated at equilibrium using isolated beam segments. Lastly, the axial force of the column can be computed by summing the shear forces acting on the ends of the beams in addition to the beam-column joint.

In conclusion, the computing procedure to determine the internal forces of lateral-force-resisting frames can be summarized in the following steps:

(1) Determine the stiffness per length and height for the beam and column;

(2) Calculate the stiffness ratio i of the beam to the column;

(3) Determine the modified factor α, quantify the degree of restraint, and calculate the D-value;

(4) Determine the column shear based on Eq. (5-6);

(5) Determine the position of the point of contraflexure in the column using Eq. (5-9);

(6) Calculate the column moments using Eqs. (5-10) and (5-11), derive the beam moments based on the equilibrium of the beam-column joint and the stiffness of the beam.

5.4.4 Internal Force under Vertical Loads

Detailed methods, such as the storey-by-storey method and the moment distribution method, are often discussed in structural analysis textbooks; thus, only brief reminders of the moment redistribution coefficient β are provided here. It is advantageous for the constriction of joints to ensure structural ductility by considering an appropriate plastic development in the beam section. In precast RC frames, $\beta=0.8—0.9$, and in cast-in-place RC frames, $\beta=0.8—$

Values of y_0 for a multi-storey frame structure with m storeys

Table 5-10

m	n	0.1	0.2	0.3	0.4	0.5	0.6	0.7	0.8	0.9	1.0	2.0	3.0	4.0	5.0
1	1	0.80	0.75	0.70	0.65	0.65	0.60	0.60	0.60	0.60	0.55	0.55	0.55	0.55	0.55
2	2	0.50	0.45	0.40	0.40	0.40	0.40	0.40	0.40	0.40	0.45	0.45	0.45	0.45	0.50
	1	1.00	0.85	0.75	0.70	0.65	0.65	0.65	0.65	0.60	0.60	0.55	0.55	0.55	0.55
3	3	0.25	0.25	0.25	0.30	0.30	0.35	0.35	0.35	0.40	0.40	0.45	0.45	0.45	0.50
	2	0.60	0.50	0.50	0.50	0.50	0.45	0.45	0.45	0.45	0.45	0.50	0.50	0.50	0.50
	1	1.15	0.90	0.80	0.75	0.75	0.70	0.70	0.65	0.65	0.65	0.55	0.55	0.55	0.55
4	4	0.10	0.15	0.20	0.25	0.30	0.30	0.30	0.35	0.30	0.40	0.40	0.45	0.45	0.45
	3	0.35	0.35	0.35	0.40	0.40	0.40	0.40	0.45	0.40	0.45	0.45	0.50	0.50	0.50
	2	0.70	0.60	0.55	0.50	0.50	0.50	0.50	0.50	0.50	0.50	0.50	0.50	0.50	0.50
	1	1.20	0.95	0.85	0.80	0.75	0.70	0.70	0.70	0.65	0.65	0.55	0.55	0.55	0.55
5	5	−0.05	0.10	0.20	0.25	0.30	0.30	0.35	0.35	0.35	0.35	0.40	0.45	0.45	0.45
	4	0.20	0.25	0.35	0.35	0.40	0.40	0.40	0.40	0.45	0.45	0.45	0.50	0.50	0.50
	3	0.45	0.40	0.45	0.45	0.45	0.45	0.45	0.45	0.45	0.45	0.50	0.50	0.50	0.50
	2	0.75	0.60	0.55	0.55	0.55	0.50	0.50	0.50	0.50	0.50	0.50	0.50	0.50	0.50
	1	1.30	1.00	0.85	0.80	0.75	0.70	0.70	0.65	0.65	0.65	0.60	0.55	0.55	0.55
6	6	−0.15	0.05	0.15	0.20	0.25	0.30	0.35	0.35	0.35	0.35	0.40	0.45	0.45	0.45
	5	0.10	0.25	0.30	0.35	0.35	0.40	0.40	0.40	0.45	0.45	0.45	0.50	0.50	0.50
	4	0.20	0.35	0.40	0.40	0.40	0.45	0.45	0.45	0.45	0.45	0.50	0.50	0.50	0.50
	3	0.50	0.45	0.45	0.45	0.45	0.45	0.45	0.50	0.50	0.50	0.50	0.50	0.50	0.50
	2	0.80	0.65	0.55	0.55	0.55	0.55	0.50	0.50	0.50	0.50	0.50	0.50	0.50	0.50
	1	1.30	1.00	0.85	0.80	0.75	0.70	0.70	0.65	0.65	0.65	0.55	0.55	0.55	0.55

Continued Table

m	n	0.1	0.2	0.3	0.4	0.5	0.6	0.7	0.8	0.9	1.0	2.0	3.0	4.0	5.0
7	7	−0.20	0.05	0.15	0.20	0.25	0.30	0.30	0.35	0.35	0.35	0.45	0.45	0.45	0.45
	6	0.05	0.20	0.30	0.35	0.35	0.40	0.40	0.40	0.40	0.45	0.45	0.50	0.50	0.50
	5	0.20	0.30	0.35	0.40	0.40	0.45	0.45	0.45	0.45	0.45	0.50	0.50	0.50	0.50
	4	0.35	0.40	0.40	0.45	0.45	0.45	0.45	0.50	0.45	0.45	0.50	0.50	0.50	0.50
	3	0.55	0.50	0.50	0.50	0.50	0.50	0.50	0.50	0.50	0.50	0.50	0.50	0.50	0.50
	2	0.80	0.65	0.60	0.55	0.55	0.55	0.50	0.50	0.50	0.50	0.50	0.50	0.50	0.50
	1	1.30	1.00	0.90	0.80	0.75	0.70	0.70	0.70	0.65	0.65	0.60	0.55	0.55	0.55
8	8	−0.20	0.05	0.15	0.20	0.25	0.30	0.30	0.35	0.35	0.35	0.45	0.45	0.45	0.45
	7	0.00	0.20	0.30	0.35	0.35	0.40	0.40	0.40	0.40	0.45	0.50	0.50	0.50	0.50
	6	0.15	0.30	0.35	0.40	0.40	0.45	0.45	0.45	0.45	0.45	0.50	0.50	0.50	0.50
	5	0.30	0.45	0.40	0.45	0.45	0.45	0.45	0.45	0.45	0.45	0.50	0.50	0.50	0.50
	4	0.40	0.45	0.45	0.45	0.45	0.50	0.50	0.50	0.50	0.50	0.50	0.50	0.50	0.50
	3	0.60	0.50	0.50	0.50	0.50	0.50	0.50	0.50	0.50	0.50	0.50	0.50	0.50	0.50
	2	0.85	0.65	0.60	0.55	0.55	0.55	0.55	0.50	0.50	0.50	0.50	0.50	0.50	0.50
	1	1.30	1.00	0.90	0.80	0.75	0.70	0.70	0.70	0.65	0.65	0.60	0.55	0.55	0.55
9	9	−0.25	0.00	0.15	0.20	0.25	0.30	0.30	0.35	0.35	0.40	0.45	0.45	0.45	0.45
	8	−0.00	0.20	0.30	0.35	0.35	0.40	0.40	0.40	0.40	0.45	0.45	0.50	0.50	0.50
	7	0.15	0.30	0.35	0.40	0.40	0.40	0.45	0.45	0.45	0.45	0.50	0.50	0.50	0.50
	6	0.25	0.35	0.40	0.40	0.45	0.45	0.45	0.45	0.50	0.50	0.50	0.50	0.50	0.50
	5	0.35	0.40	0.45	0.45	0.45	0.50	0.50	0.50	0.50	0.50	0.50	0.50	0.50	0.50
	4	0.45	0.45	0.45	0.45	0.50	0.50	0.50	0.50	0.50	0.50	0.50	0.50	0.50	0.50
	3	0.60	0.55	0.50	0.50	0.50	0.50	0.55	0.50	0.50	0.50	0.50	0.50	0.50	0.50
	2	0.85	0.65	0.60	0.55	0.55	0.55	0.55	0.50	0.50	0.50	0.50	0.50	0.50	0.50
	1	1.35	1.00	0.90	0.80	0.75	0.75	0.70	0.70	0.65	0.65	0.60	0.55	0.55	0.55

Continued Table

m	n	0.1	0.2	0.3	0.4	0.5	0.6	0.7	0.8	0.9	1.0	2.0	3.0	4.0	5.0
10	10	−0.25	0.00	0.15	0.20	0.25	0.30	0.30	0.35	0.35	0.40	0.45	0.45	0.45	0.45
	9	−0.05	0.20	0.30	0.35	0.35	0.40	0.40	0.40	0.40	0.45	0.45	0.50	0.50	0.50
	8	−0.10	0.30	0.35	0.40	0.40	0.40	0.45	0.45	0.45	0.45	0.50	0.50	0.50	0.50
	7	0.20	0.35	0.40	0.40	0.45	0.45	0.45	0.45	0.45	0.50	0.50	0.50	0.50	0.50
	6	0.30	0.40	0.40	0.45	0.45	0.45	0.45	0.50	0.50	0.50	0.50	0.50	0.50	0.50
	5	0.40	0.45	0.45	0.45	0.45	0.45	0.50	0.50	0.50	0.50	0.50	0.50	0.50	0.50
	4	0.50	0.45	0.45	0.45	0.50	0.50	0.50	0.50	0.50	0.50	0.50	0.50	0.50	0.50
	3	0.60	0.55	0.50	0.50	0.50	0.50	0.55	0.50	0.50	0.50	0.50	0.50	0.50	0.50
	2	0.85	0.65	0.55	0.55	0.55	0.55	0.55	0.50	0.50	0.50	0.50	0.50	0.50	0.50
	1	1.35	1.00	0.80	0.80	0.75	0.75	0.70	0.70	0.65	0.65	0.60	0.55	0.55	0.55
11	11	−0.25	0.00	0.15	0.20	0.25	0.30	0.30	0.30	0.35	0.35	0.45	0.45	0.45	0.45
	10	0.05	0.20	0.25	0.30	0.35	0.40	0.40	0.40	0.40	0.45	0.45	0.50	0.50	0.50
	9	0.10	0.30	0.35	0.40	0.40	0.45	0.45	0.45	0.45	0.45	0.50	0.50	0.50	0.50
	8	0.20	0.35	0.40	0.40	0.45	0.45	0.45	0.45	0.45	0.50	0.50	0.50	0.50	0.50
	7	0.25	0.40	0.40	0.45	0.45	0.45	0.45	0.50	0.50	0.50	0.50	0.50	0.50	0.50
	6	0.30	0.40	0.45	0.45	0.45	0.50	0.50	0.50	0.50	0.50	0.50	0.50	0.50	0.50
	5	0.40	0.45	0.45	0.45	0.45	0.50	0.50	0.50	0.50	0.50	0.50	0.50	0.50	0.50
	4	0.50	0.50	0.45	0.50	0.50	0.50	0.50	0.50	0.50	0.50	0.50	0.50	0.50	0.50
	3	0.65	0.55	0.50	0.55	0.55	0.55	0.55	0.50	0.50	0.50	0.50	0.50	0.50	0.50
	2	0.85	0.65	0.60	0.55	0.55	0.55	0.55	0.50	0.50	0.50	0.50	0.50	0.50	0.50
	1	1.35	1.05	0.90	0.80	0.75	0.75	0.70	0.70	0.65	0.65	0.60	0.55	0.55	0.55

Continued Table

m	n	0.1	0.2	0.3	0.4	0.5	0.6	0.7	0.8	0.9	1.0	2.0	3.0	4.0	5.0
12+	m−1	−0.30	0.00	0.15	0.20	0.25	0.30	0.30	0.30	0.35	0.35	0.45	0.45	0.45	0.45
	m−2	−0.10	0.20	0.25	0.30	0.35	0.40	0.40	0.40	0.40	0.40	0.45	0.45	0.45	0.50
	m−3	0.05	0.25	0.35	0.40	0.40	0.40	0.40	0.45	0.45	0.45	0.50	0.50	0.50	0.50
	m−4	0.15	0.30	0.40	0.40	0.45	0.45	0.45	0.45	0.45	0.45	0.50	0.50	0.50	0.50
	m−5	0.25	0.35	0.40	0.45	0.45	0.45	0.45	0.45	0.50	0.50	0.50	0.50	0.50	0.50
	m−6	0.30	0.40	0.40	0.45	0.45	0.50	0.50	0.50	0.50	0.50	0.50	0.50	0.50	0.50
	Mid	0.45	0.45	0.45	0.45	0.50	0.50	0.50	0.50	0.50	0.50	0.50	0.50	0.50	0.50
	4	0.55	0.50	0.50	0.50	0.50	0.50	0.50	0.50	0.50	0.50	0.50	0.50	0.50	0.50
	3	0.65	0.55	0.50	0.50	0.50	0.50	0.50	0.50	0.50	0.50	0.50	0.50	0.50	0.50
	2	0.70	0.70	0.60	0.55	0.55	0.55	0.55	0.50	0.50	0.50	0.60	0.55	0.55	0.55
	1	1.35	1.05	0.90	0.80	0.70	0.75	0.70	0.70	0.70	0.65	0.60	0.55	0.55	0.55

Note: The values of the first line represent the ratio of the stiffness of beams and columns.

Adjustment coefficient y_1 for different beam stiffness ratios between two adjacent floors   Table 5-11

K \ I	0.1	0.2	0.3	0.4	0.5	0.6	0.7	0.8	0.9	1.0	2.0	3.0	4.0	5.0
0.4	0.55	0.40	0.30	0.25	0.20	0.20	0.20	0.15	0.15	0.15	0.05	0.05	0.05	0.05
0.5	0.45	0.30	0.20	0.20	0.15	0.15	0.15	0.10	0.10	0.10	0.05	0.05	0.05	0.05
0.6	0.30	0.20	0.15	0.15	0.10	0.10	0.10	0.10	0.05	0.05	0.05	0.05	0	0
0.7	0.20	0.15	0.10	0.10	0.10	0.10	0.10	0.05	0.05	0.05	0	0	0	0
0.8	0.15	0.10	0.05	0.05	0.05	0.05	0.05	0.05	0.05	0	0	0	0	0
0.9	0.05	0.05	0.05	0.05	0	0	0	0	0	0	0	0	0	0

Adjustment coefficients y_2 and y_3 for different height differences between two adjacent storeys Table 5-12

a_2	a_3 \ i	0.1	0.2	0.3	0.4	0.5	0.6	0.7	0.8	0.9	1.0	2.0	3.0	4.0	5.0
2.0		0.25	0.15	0.15	0.10	0.10	0.10	0.10	0.10	0.05	0.05	0.05	0.05	0.0	0.0
1.8		0.20	0.15	0.10	0.10	0.10	0.05	0.05	0.05	0.05	0.05	0.05	0.0	0.0	0.0
1.6	0.4	0.15	0.10	0.10	0.05	0.05	0.05	0.05	0.05	0.05	0.05	0.0	0.0	0.0	0.0
1.4	0.6	0.10	0.05	0.05	0.05	0.05	0.05	0.05	0.0	0.05	0.0	0.0	0.0	0.0	0.0
1.2	0.8	0.05	0.05	0.05	0.0	0.0	0.0	0.0	0.0	0.0	0.0	0.0	0.0	0.0	0.0
1.0	1.0	0.0	0.0	0.0	0.0	0.0	0.0	0.0	0.0	0.0	0.0	0.0	0.0	0.0	0.0
0.8	1.2	-0.05	-0.05	-0.05	-0.05	-0.05	-0.05	-0.05	-0.05	-0.05	0.0	0.0	0.0	0.0	0.0
0.6	1.4	-0.10	-0.05	-0.05	-0.05	-0.05	-0.05	-0.05	-0.05	-0.05	-0.05	0.0	0.0	0.0	0.0
	1.6	-0.15	-0.10	-0.10	-0.10	-0.10	-0.05	-0.05	-0.05	-0.05	-0.05	-0.05	0.0	0.0	0.0
0.4	1.8	-0.20	-0.15	-0.10	-0.10	-0.10	-0.10	-0.10	-0.05	-0.05	-0.05	-0.05	0.0	0.0	0.0
	2.0	-0.25	-0.15	-0.15	-0.10	-0.10	-0.10	-0.10	-0.10	-0.05	-0.05	-0.05	-0.05	0.0	0.0

0.9. Furthermore, the moment at the mid-span of the beam must increase accordingly.

5.4.5 Loads Combination

In this section, two main issues are discussed: (1) load combinations based on reliability design and (2) member design based on seismic adjustment and internal force amplification factors.

Before combining the loads, it is important to examine the critical sections of the frame members. The critical design sections of the beams in a structure's frame are adjacent to the beam-column joint at the two ends and the mid-span, while for the columns, the upper section and the lower section near the joint are the most critical.

In the seismic design of multi-storey and high-rise RC buildings in China, the structures must be designed to withstand the unfavorable combination of designed loads. The following basic combinations can be considered in accordance with the ***Code for Seismic Design of Buildings*** (GB 50011—2010) Version 2016 and ***General Code for Seismic Precaution of Buildings and Municipal Engineering*** (GB 55002—2021):

$$S = \gamma_G S_{GE} + \gamma_{EH} S_{EHk} \quad (5\text{-}12)$$
$$S = \gamma_G S_{GE} + \psi_w \gamma_w S_{wk} + \gamma_{EH} S_{EHk} \quad (5\text{-}13)$$
$$S = \gamma_G S_{Gk} + \gamma_Q S_{Qk} \quad (5\text{-}14)$$

where,

γ_G——the load factor for a dead load and no less than 1.3;

γ_{EH}——the load factor for a horizontal seismic action, i.e., 1.4;

γ_Q——the load factor for a variable load, i.e., 1.5;

γ_w——the load factor for a wind load, i.e., 1.5;

ψ_w——the coefficient of the wind load combination, which is 0 for ordinary structures and 0.2 for high-rise buildings when the wind load is the control case;

S_{GE}——the representative value of the gravity load effect;

S_{EHk}——the characteristic value of the horizontal seismic action;

S_{wk}——the characteristic value of the wind load effect;

S_{Gk}——the representative value of the dead load;

S_{Qk}——the characteristic value of the variable load.

For certain structures (e.g., isolated structures), when vertical seismic action is considered:

$$S = \gamma_G S_{GE} + \gamma_{EH} S_{EHk} + \gamma_{EV} S_{EVk} \quad (5\text{-}15)$$

where,

S_{EVk}——the characteristic value of the vertical seismic action, which should be multiplied by the relevant amplification factor or adjustment factor;

γ_{EV}——the load factor for the vertical seismic action, which is 1.4 when only the vertical seismic action is considered or both the horizontal and vertical seismic actions are considered with the primary action vertical and 0.5 when both the horizontal and vertical seismic actions are considered with the primary action horizontal.

Based on the above combinations, the structural members must be designed to withstand the most unfavorable combination:

$$S \leq R/\gamma_{RE} \quad (5\text{-}16)$$

where,

S——the design value of the combinations for a structural member, including the moment, shear and axial force;

R——the design value of the strength resistance of the structural member;

γ_{RE}——the seismic adjustment factor.

Furthermore, different members are assigned different γ_{RE} values, which are related to (1) the materials, (2) the member types, (3) the control conditions (i.e., strength and stability) for steel members and (4) the stress conditions (i.e., shear, bending, eccentric compression and eccentric tension) for masonry and concrete members. However, γ_{RE} will never exceed 1.0 because it represents a downscale trend due to the uncertainty of seismic events. When only the vertical seismic action is considered, the seismic adjusting factor should be 1.0 for all structural members.

5.4.6 Cross-Section Design

The second issue is related to the design of member cross-sections. Because comparatively thorough and systemic design methods and rules are introduced in the ***Code for Seismic Design of Buildings*** (GB 50011—2010) Version 2016, a civil engineer can obtain the necessary guidance by carefully learning the code. Thus, a simple checklist (Table 5-13) is provided, more attention is given to certain essentials and principles later.

Checklist for the member design process   Table 5-13

Object type	Item
Beam	Bending moment capacity at critical cross-sections
Beam	Shear-compression ratio
Beam	Adjustment of the shear design value with η_{vb}
Beam	Shear capacity
Column	(Eccentric) Compression capacity
Column	Axial force ratio
Column	Adjustment of the bending moment design value with η_c
Column	Shear-compression ratio
Column	Adjustment of the shear design value with η_{vc}
Column	Shear capacity
Joint	Shear-compression ratio
Joint	Shear capacity
Joint	Adjustment of the shear design value in the core zone with η_{jb}

Furthermore, several topics should be expanded upon:

(1) The "three-strong and three-weak" seismic design principle

To avoid global collapse and the creation of additional plastic hinges, the first principle, i.e., "strong column and weak beam", is recommended. Because flexure failure is more favorable than shear failure in term of ductility, the second principle, i.e., "strong shear and weak bending", is widely suitable for different members, and severe damage and arduous repairs after earthquakes inspire the third law, i.e., "strong joint and weak member".

In reality, these three principles can be executed through internal force amplification factors η_i as follows:

$$V = \eta_{vb}(M_b^l + M_b^r)/l_n + V_{Gb} \quad (5\text{-}17)$$

$$\sum M_c = \eta_c \sum M_b \quad (5\text{-}18)$$

$$V = \eta_{vc}(M_c^t + M_c^b)/H_n \quad (5\text{-}19)$$

$$V = \frac{\eta_{jb} \sum M_b}{h_{b0} - \alpha'_s}\left(1 - \frac{h_{b0} - \alpha'_s}{H_c - h_b}\right) \quad (5\text{-}20)$$

where,

V——the design value of combined shear force on the end section;

η_{vb}, η_{vc}——the enhancement coefficients for the shear force at beam and column ends, respectively;

M_b^l, M_b^r——the design values of bending moments on the right and left sides of the beam, respectively;

l_n——the clear span of the beam;

V_{Gb}——the design value of shear force of

beam;

M_c —— the combined design value on the column end sections at one joint;

M_c^t, M_c^b —— the design value of bending moments on the top and bottom sides of the beam, respectively;

H_n —— the clear height of the column;

$\sum M_b$ —— the combined design value on the beam end sections at one joint;

H_c —— the column height;

h_b —— the beam height;

h_{b0} —— the effective beam height;

η_{jb} —— the enhancement coefficient for beam-column joint;

α_s' —— the distance between the resultant force of the compression reinforcement and the section edge.

(2) Shear-compression ratio

Shear stress is unfavorable for ductility, and based on previous studies, even stirrups will fail to protect the shear section when the cross-section is too small. Thus, to control the minimum scale of a cross-section, Eq. (5-21) can be used to define the largest shear stress:

$$V \leq \frac{1}{\gamma_{RE}}(kf_c bh_0) \quad (5\text{-}21)$$

where,

V —— the design value of combined shear force on the end section;

γ_{RE} —— the seismic adjustment factor;

k —— the coefficient depending on component shear span ratio;

f_c —— the compressive strength of the concrete;

b, h_0 —— the width and effective height of the cross-section, respectively.

(3) Axial force ratio

Similar to the shear-compression ratio for columns, the axial force is a defined parameter because of its adverse effect on ductility. Specifically, the maximum values are listed in Table 5-14 for reference.

Axial force ratio limits   Table 5-14

Structure type	Seismic grade			
	1	2	3	4
Frame structures	0.65	0.75	0.85	0.90
Frame-wall, slab-column-wall and tube structures	0.75	0.85	0.90	0.95
Frame-support-wall structures	0.60	0.70	N/A	

5.4.7 Lateral Displacement Check

Drift or inter-storey displacement should be limited (which might cause non-structural damage, human discomfort and secondary stresses in the main structure) to prevent structural collapse. The floor drift is normally specified at the elastic design level, although it is greater under maximum earthquake action conditions. The *Code for Seismic Design of Buildings* (GB 50011—2010) Version 2016 specifies that the values vary between 0.002 and 0.008 times that of the storey height, with 0.005 being commonly used.

Additionally, the code specifies that the elastic drift must be checked for all frame structures in a frequent earthquake intensity level; however, for certain framed buildings, the plastic drift should be checked for a rare earthquake intensity level.

(1) Checking elastic floor drifts under frequently occurred earthquake

When checking the lateral deflection in the elastic state, the elastically calculated deflection can be used. Under a frequently occurring earthquake intensity level, the elastic drift of the floor of frame structures must meet the following requirement:

$$\Delta u_e \leq [\theta_e]h \quad (5\text{-}22)$$

where,

Δu_e —— the maximum elastic floor drift induced by the characteristic lateral seismic loading under a frequent earthquake intensity level;

except for high-rise buildings with predominant flexural deformation, the flexural deformation should not be deduced from the calculated drift, and the torsion deformation should be considered; when performing the calculation, all load factors should be assumed to be 1.0, and the elastic stiffness of the members can be used;

$[\theta_e]$——the upper bound of the elastic floor drift rotation, which is 1/550 for RC frames;

h——the storey height.

(2) Checking plastic floor drifts under rarely occurred earthquake

Similar to the elastic floor drift, the following equation provides the requirement for the plastic floor drift of a structure:

$$\Delta u_P \leqslant [\theta_P] h \qquad (5\text{-}23)$$

where,

$\Delta u_P, [\theta_P]$——the plastic deformation and plastic rotation.

The upper bound of the elastoplastic floor drift is 1/50 for RC frames. When the axial force ratio of the framed column is less than 0.4, the limit can be increased by 10%. The recommended values can be referred to in the codes for the respective seismic grade and stirrup types in the columns.

Furthermore, a simplified equation to calculate the elastoplastic floor drift is as follows:

$$\Delta u_P \leqslant \eta_P \Delta u_e \qquad (5\text{-}24)$$

where,

η_P——the amplification coefficient for the elastoplastic floor drift.

Chapter 6
Seismic Design of Masonry Building Structures

6.1 Introduction

6.1.1 Masonry Structures

Masonry structures are widely used in China. The popularity of masonry construction is due to the relatively simple construction process and economical construction. Masonry has been used in many types of buildings, such as residential buildings, schools and hospitals. The current national code applies to multi-storey buildings bearing with masonry, such as common bricks (including fired, autoclaved and concrete common bricks), perforated bricks (including fired and concrete perforated bricks) and small hollow blocks as well as masonry buildings with reinforced concrete (RC) frames for the ground floors.

Multi-storey masonry buildings refer to a type of structure that uses masonry walls as vertical bearing members and RC walls or walls using other materials as horizontal load-bearing elements (floor and roof). Masonry buildings with RC frames for the ground floors refer to a special type of structure that mainly exists in China, which uses RC frames for the ground floor (or two storeys from the ground floor) and masonry systems for the upper storeys. It is vertically irregular, and the lower storeys may suffer considerable damage during earthquakes. Generally, masonry structures are primarily constructed with brittle materials, which have low tensile, shear and bending strengths. After the Tangshan Earthquake in 1976, research has been conducted in China to promote seismic design and construction details, which are reflected in the *Gode for Seismic Design of Industrial and Civil Buildings* (TJ 11—78) and *Code for Seismic Design of Buildings* (GB 50011—2010) Version 2016.

6.1.2 Ductile Design

To have masonry structures in high seismic regions capable of significant inelastic behavior without failure. The two primary measures used in masonry structures to prevent brittle failure are confined boundary elements and reinforcement. Confined boundary elements, which consist of constructional columns, core columns and ring beams, confine the masonry walls along the edges. Constructional columns are commonly placed at corners, wall intersections as well as the vertical edges of door or window openings. Columns should be well connected to the horizontal bonding elements by reinforcements. Ring beams, which are commonly placed at the floor and roof level, can significantly improve the seismic capacity of the entire structure. The reinforcement consists of deformed reinforcing bars or joint reinforcements. Deformed reinforcing bars are placed vertically in the cells of hollow elements, horizontally in courses of bond-beam units, or vertically and horizontally between layers of solid units. They can be encased in concrete or grout. The joint reinforcement is placed in bed joints (horizontal) of masonry walls and surrounded by mortar.

6.2 Structural Configuration and Systems

6.2.1 Structural Configuration

Generally, during the conceptual design process, the layout of the plan and the elevation of buildings should be regular and symmetrical. Some detailed considerations are described below.

1) Multi-storey masonry buildings

(1) Priority shall be given to the use of structural systems with transverse load-bearing walls or both transverse and longitudinal load-bearing walls (Fig. 6-1).

(2) The distribution of longitudinal and transverse walls should be uniform and symmetrical, aligned in-plane, and shall be continued in elevation. The walls between the windows on the same axis should be distributed uniformly.

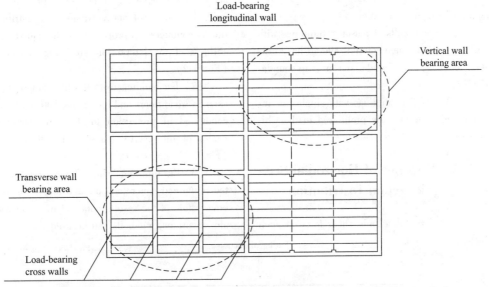

Fig. 6-1 Diagram of a masonry structure with a vertical load-bearing wall and a load-bearing cross wall

(3) A seismic joint shall be built in intensities 8 and 9, as well as in one of the following instances. Walls shall be installed on both sides of the joint, and the joint gap should be between 70 and 100 mm. The difference in heights on the elevation of buildings exceeds 6 m. In buildings with staggered storeys and with a relatively large difference (exceeds 1/4 storey height) between floor levels in elevation, the stiffness and mass of different parts of the structure are sharply different.

(4) Staircases should not be installed at the end of a building or a corner.

(5) The installation of chimneys, ventilation ducts and waste ducts in a building shall not weaken the wall. When the wall is weakened, strengthening measures shall be taken. Chimneys adhered to the wall without vertical reinforcement or chimneys on the roof should not be used.

(6) Precast RC eaves without anchorage should not be used.

2) Multi-storey masonry buildings with the first one/two frame-supported storeys

(1) The masonry shear wall at the top should be aligned with the beams of the frame or shear wall in the lower framed part.

(2) In the lower framed storeys, certain shear walls shall be symmetrically oriented along with the longitudinal and transverse directions, and shear walls should be uniform and symmetrical basically.

(3) For the masonry buildings with frame-supported first one storey, the ratio of lateral drift stiffness of the second storey to that of the first storey shall not be larger than 2.5 and 2.0 in intensities 7 and 8, respectively.

(4) For the masonry buildings with frame-supported first two storeys, the ratio of lateral drift stiffness of the third storey to that of the second storey shall not be larger than 2.0 and 1.5 in intensities 7 and 8, respectively. And the stiffness of the first one should be close to that of the second storey.

(5) Shear walls of masonry buildings with lower framed storeys should have raft-style footings, pile foundations or strip foundations.

3) Multi-storey masonry buildings with inner frames

(1) Buildings shall have a rectangular design and a regular elevation.

(2) Shear walls of masonry buildings with multi-row inner frame columns should have raft-style footings, pile foundations or strip foundations.

(3) Masonry buildings with single-row interior frame columns are prohibited in the Chinese code.

6.2.2 Principle of Horizontal Seismic Action Distribution

The horizontal seismic shear force at each floor level of a structure can be distributed among the lateral force-bearing structural members (such as the walls, columns and shear walls) according to the following principles:

(1) For buildings with rigid diaphragms, such as cast-in-place and monolithic-precast reinforced concrete floors and roofs, the distribution is conducted in proportion to the equivalent stiffness of the lateral force-resisting members (Fig. 6-2).

(2) For buildings with flexible diaphragms, such as wood floors and roofs, the distribution is conducted in proportion to the representative values of the gravity loads on the floor area (Fig. 6-3).

(3) For buildings with ordinary prefabricated reinforced concrete floors and roofs, the average value of the results obtained from the two aforementioned methods can be used.

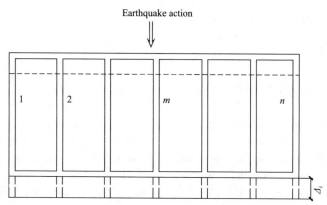

Fig. 6-2 Deformation of a rigid slab

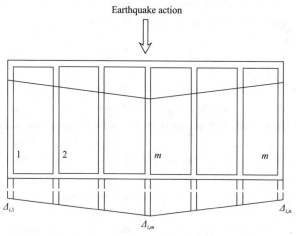

Fig. 6-3 Deformation of a flexible slab

6.3 Basic Seismic Design for Multi-Storey Masonry Buildings

The base shear method may be used in the seismic design of multi-storey masonry buildings, as well as masonry buildings with RC frames on the ground floors and seismic walls. The seismic force calculation, as shown in Fig. 6-4, begins with a simplification of the structural model. The procedure includes the following components:

(1) Estimate the equivalent weight of each floor G_i

(2) Calculate the total equivalent weight of the building

$$G_{eq} = \begin{cases} G_i, & n = 1 \\ 0.85 \sum_{i=1}^{n} G_i, & n > 1 \end{cases}$$

(3) Calculate the base shear seismic force

$$F_{Ek} = \alpha_{max} \cdot G_{eq}$$

(4) Distribute the seismic force to each floor (Fig. 6-4)

$$F_{Ek} = \frac{G_i H_i}{\sum_{j=1}^{n} G_j H_j}$$

(5) Compute the shear force at each floor, which is indicated in Fig. 6-5

The earthquake action effect at the top of a building should be adjusted if necessary. A section seismic capacity check is then performed on the wall sections with a greater bearing area or less vertical stress.

6.3.1 Equivalent Wall Stiffness

In the distribution of seismic shear forces and the cross-section check, the equivalent wall stiffness for the lateral force resistance of a storey can be determined using the following principles (Fig. 6-6).

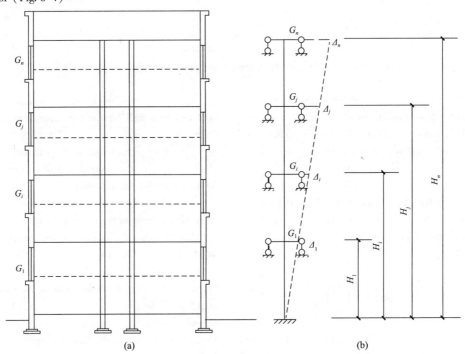

Fig. 6-4 Structural model for the seismic analysis of a multi-storey masonry building
(a) Multi-storey masonry structure; (b) Computing model

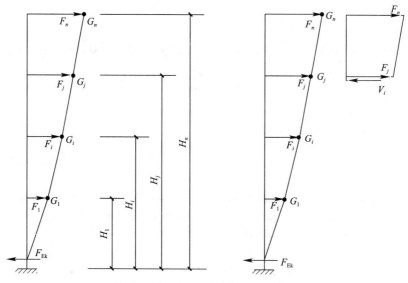

Fig. 6-5 Shear force calculation at each floor

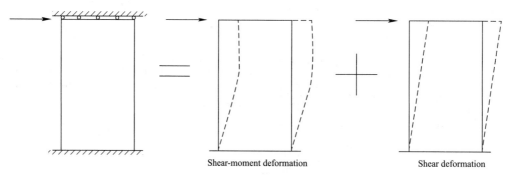

Fig. 6-6 Horizontal deformation of a wall

(1) Only the shear deformation should be considered because the height-width ratio of the wall section is less than 1.0, and the wall stiffness can be calculated as follows:

$$K_{im} = \frac{GA_{im}}{\zeta hi} \quad (6\text{-}1)$$

(2) Both the bending and shear deformations should be considered when the height-width ratio of the wall section is between 1.0 and 4.0:

$$K_{im} = \frac{E_i t_{im} b_{im}}{hi\left[3 + \left(\dfrac{hi}{b_{im}}\right)^2\right]} \quad (6\text{-}2)$$

where,

G, E——the shear modulus and elastic modulus of the masonry material, respectively; $G = 0.4E$;

t_{im}, b_{im}, A_{im}——the thickness, width, and area of the m-th wall for the i-th storey, respectively;

h_i——the height of the i-th storey;

K_{im}——the equivalent wall stiffness of the m-th wall for the i-th storey;

ζ——the non-uniformity coefficient of the shear stress in the cross-section.

The stiffness of the wall section can be ignored when the height-width ratio of the wall section exceeds 4.0. The height-width ratio of the wall section refers to the ratio of the storey height to the length of the wall. For small wall sections at the edge of door or window openings, it refers to the ratio of the height of the opening to the width of the wall at the edge of the opening (Fig. 6-7).

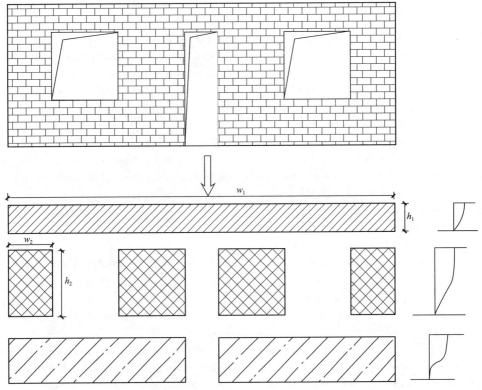

Fig. 6-7 Lateral stiffness of a wall with openings

6.3.2 Distribution of Shear Force on Floors

According to the stiffness of the slab, it can be divided into three types: rigid diaphragms, flexible diaphragms and semi-rigid diaphragms.

1) Rigid diaphragms

The rigid floor refers to the floor with infinite in-plane stiffness, which is assumed to have rigid zone deformation without any in-plane deformation under the horizontal earthquake effect. The rigid floor mainly refers to the cast-in-place and precast RC floor and roof (which have strong cast-in-place reinforced surface layer without large opening) that comply with the seismic code. Because the torsional effects can be ignored in masonry structure, the lateral displacement of the walls which support the rigid floor is equal.

The lateral stiffness of m-th wall is K_{im}, the relative lateral deformation is Δ_{im}, that the shear force distributed to the m-th wall can be calculated by:

$$V_{im} = K_{im}\Delta_{im}$$

Because the floor is rigid, the $\Delta_{im} = \Delta_i$ ($m = 1,2,\cdots,l$), that earthquake shear force of the m-th wall is:

$$V_{im} = \frac{K_{im}}{\sum_{m=1}^{l} K_{im}} V_i = \frac{K_{im}}{K_i} V_i$$

where,

$K_i = \sum_{m=1}^{l} K_{im}$——the total lateral stiffness of i-th floor.

When the height-width ratio of mostly walls is less than 1.0 and does not be influenced by opening, the lateral stiffness of each wall section can be simplified as the shear stiffness, and the shear distribution can be simplified as the distribution according to the shear section area of the wall section:

$$V_{im} = \frac{\dfrac{G_i A_{im}}{\xi h_i}}{\sum_{m=1}^{l} \dfrac{G_i A_{im}}{\xi h_i}} V_i = \frac{A_{im}}{\sum_{m=1}^{l} A_{im}} V_i = \frac{A_{im}}{A_i} V_i$$

2) Flexible diaphragms

The flexible floor refers to the floor without in-plane stiffness, so that each wall is assumed to deform freely and is not constrained by the floor under the horizontal earthquake effect. The flexible floor mainly refers to the wood floor and prefabricated RC floor with lower integrity, which usually does not comply with the seismic concept design requirements about regularity. Because the deformation of transverse walls is not continuous, the floor can be approximated as a multi-span simply supported beam that simply supported on the top of each wall. And the seismic action of each wall section is distributed by the representative gravity load of the subordinate area of the wall.

In this case, the deformation of each wall is related to the inertial force, and it is considered that the seismic shear force borne by each wall is proportional to the representative value of the gravity load borne by the wall, that:

$$V_{im} = \frac{G_{im}}{G_i} V_i$$

where,

G_{im}——the representative gravity load of the subordinate area of the m-th wall in the i-th floor;

G_i——the representative gravity load of the i-th floor.

When the gravity of each floor is evenly distributed along the plane:

$$G_{im} = \gamma_i S_{im}, \quad G_i = \gamma_i S_i$$

where,

γ_i——the representative gravity load of the unit area in i-th floor;

S_{im}——subordinate area of the m-th wall;

S_i——the total area of the i-th floor.

$$V_{im} = \frac{S_{im}}{S_i} V_i$$

3) Semi-rigid diaphragms

The semi-rigid floor refers to the floor of which the in-plane stiffness is between rigid floor and flexible floor, such as ordinary prefabricated RC floor. When using semi-rigid floor, it is complicated to calculate the seismic shear force of each wall, so the average value of the results obtained from the above-mention two methods of distribution may be used:

$$V_{im} = \frac{1}{2}\left(\frac{K_{im}}{K_i} + \frac{S_{im}}{S_i}\right) V_i$$

In actual engineering, when the height-width ratio of mostly walls are smaller than 1, it can be simplified as:

$$V_{im} = \frac{1}{2}\left(\frac{A_{im}}{A_i} + \frac{S_{im}}{S_i}\right) V_i$$

6.3.3 Seismic Shear Bearing Capacity of Walls

1) Walls with common bricks or perforated bricks

The seismic shear bearing capacity of walls with common brick and perforated brick can be determined using the following approach:

$$V \leqslant \frac{f_{vE} A}{\gamma_{RE}} \quad (6\text{-}3)$$

where,

V——the design value of the shear force for a wall;

A——the wall cross-section area (area of wall cross-section at 1/2 storey height); the perforated brick should use the gross sectional area;

γ_{RE}——the seismic adjustment coefficient of the bearing capacity; for self-supporting walls, it is 0.75; for seismic walls with constructional columns and core columns at both ends, it is 0.9; and for other seismic walls, it is 1.0;

f_{vE}——the design value of the seismic shear strength of the masonry damaged along the stepped cross-section of the brick.

$$f_{vE} = \xi_N f_v \quad (6\text{-}4)$$

f_v——the design value of the shear strength of the masonry material for a non-seismic design;

ξ_N——the normal stress influence coefficient of the seismic shear strength of the masonry material, which can be obtained from Table 6-1.

Normal stress influence coefficient for masonry wall strength ξ_N   Table 6-1

Type of masonry walls	σ_0/f_v							
	0.0	1.0	3.0	5.0	7.0	10.0	12.0	≥16.0
Common brick and perforated brick	0.80	0.99	1.25	1.47	1.65	1.90	2.05	—
Small block	—	1.23	1.69	2.15	2.57	3.02	3.32	3.92

Note: σ_0 is the mean compressive stress of the masonry section corresponding to the representative value of the gravity load.

2) Brick walls with horizontal reinforcement

The seismic shear bearing capacity of walls with a horizontal reinforcement can be determined using the following approach:

$$V \leqslant \frac{1}{\gamma_{RE}}(f_{vE}A + \xi_s f_y A_{sh}) \quad (6-5)$$

where,

f_y —— the design value of the tensile strength of the horizontal reinforcement;

A_{sh} —— the total horizontal reinforcement area of the vertical sections in the wall of the storey, for which the reinforcement percentage should be at larger than 0.07% but less than 0.17%;

ξ_s —— the reinforcement participation factor, which can be selected using Table 6-2.

3) Walls with small blocks

The seismic load-bearing capacity for the cross-section of walls with small concrete blocks can be determined as follows:

$$V \leqslant \frac{1}{\gamma_{RE}}[f_{vE}A + (0.3f_t A_c + 0.05 f_y A_s)\zeta_c] \quad (6-6)$$

where,

f_t —— the design value of the tensile strength of the concrete center of the core column;

A_c —— the total cross-sectional area of the core column;

A_s —— the total cross-sectional area of the steel reinforcement for the core column;

ζ_c —— the participation factor of the core column, which can be selected from Table 6-3.

Reinforcement participation service factor   Table 6-2

Height-width ratio of the wall	0.4	0.6	0.8	1.0	1.2
ξ_s	0.10	0.12	0.14	0.15	0.12

Participation service coefficient of the core column   Table 6-3

Filled opening ratio ρ	$\rho<0.15$	$0.15 \leqslant \rho < 0.25$	$0.25 \leqslant \rho < 0.5$	$\rho \geqslant 0.5$
ζ_c	0.0	1.0	1.10	1.15

6.4 Basic Seismic Design of Multi-Storey Masonry Buildings with the First One/Two Frame-Supported Storeys

The base shear method may be used for seismic calculations of brick buildings with a frame system at the bottom. The seismic shear force design values in the longitudinal and transversal directions of the frame system in masonry buildings should be multiplied by a magnification factor in the range of 1.2—1.5 based on the ratio of the lateral stiffness of the upper storey to that of the frame system storey. The seismic effects of the frame columns in masonry buildings can be determined using the following method:

1) The seismic shear force design values of the frame columns are determined by the ratio of the column lateral stiffness to the stiffness of all lateral force-resisting members. The effective lateral stiffness cannot be reduced for the frame but can be reduced to 30% for reinforced concrete walls and 20% for clay brick walls.

2) An additional axial force induced by the seismic overturning moment should be considered when calculating the frame column axial forces (Fig. 6-8). The seismic overturning moment of the elements along all axes can be determined by the ratio of the lateral stiffness of the shear walls and the frame in the first storey. Additional axial forces and shear forces can be determined as follows:

$$N_f = V_w H_f / l \quad (6\text{-}7)$$
$$V_f = V_w \quad (6\text{-}8)$$

where,

V_w——the shear force design value of the wall;

N_f——the additional axial force design value of the frame column;

V_f——the additional shear force design value of the frame column;

H_f, l——the storey height and span of the frame, respectively.

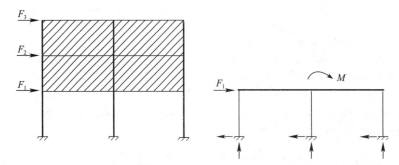

Fig. 6-8 Ground floor frame overturning evaluation

6.5 Limitations and Reasons

In general, the total height and number of storeys for masonry buildings should not exceed those specified in Table 6-4. However, for certain buildings, such as hospitals and schools, the values in Table 6-4 should be reduced by 3 m and/or one storey; for buildings with extremely few transverse bearing walls in each storey, these values should be further reduced based on specific conditions. The storey height of masonry buildings should not be larger than 3.6 m. The maximum ratio of the total height and total width of a multi-storey masonry building should follow the requirements in Table 6-5. The total width of buildings with single-side corridors does not include the corridor width. The height-width ratio should be reduced if the architectural plan is similar to a square. The spacing between the seismic transverse shear walls of multi-storey masonry buildings should not exceed the values listed in Table 6-6. The local dimension limitation of multi-storey masonry buildings should meet the requirements enumerated in Table 6-7.

Table 6-4 Number of storeys and total height limitation for buildings (m)

Building category		Minimum thickness of seismic walls (mm)	VI (6) 0.05 g		VII (7) 0.10 g		VII (7) 0.15 g		VIII (8) 0.20 g		VIII (8) 0.30 g		IX (9) 0.40 g	
			Height	Number of storeys	Height	Number of storeys	Height	Number of storeys	Height	Number of storeys	Height	Number of storeys	Height	Number of storeys
Multi-storey masonry buildings	Ordinary brick	240	21	7	21	7	21	7	18	6	15	5	12	4
	Perforated brick	240	21	7	21	7	18	6	18	6	15	5	9	3
	Perforated brick	190	21	7	18	6	15	5	15	5	12	4	—	—
	Small block	190	21	7	21	7	18	6	18	6	15	5	9	3
Masonry buildings with RC frames on ground floors and seismic walls	Ordinary rick and perforated brick	240	22	7	22	7	19	6	16	5	—	—	—	—
	Perforated brick	190	22	7	19	6	16	5	13	4	—	—	—	—
	Small block	190	22	7	22	7	19	6	16	5	—	—	—	—

Maximum height-width ratio for buildings Table 6-5

Intensity	VI(6)	VII(7)	VIII(8)	IX(9)
Maximum height-width ratio	2.5	2.5	2.0	1.5

Spacing of transverse walls (m) Table 6-6

Building category		Intensity			
		VI(6)	VII(7)	VIII(8)	IX(9)
Multi-storey masonry buildings	Cast-in-situ or prefabricated integral RC floor and roof slabs	15	15	11	7
	Prefabricated RC floor and roof slabs	11	11	9	4
	Wooden roof	9	9	4	—
Bottom frame-seismic wall masonry buildings	Upper storeys	Same as for multi-storey masonry buildings			—
	Lower storeys	18	15	11	—

Dimensional limit (m) Table 6-7

Position	Intensity			
	VI(6)	VII(7)	VIII(8)	IX(9)
The minimum width between load-bearing windows	1.0	1.0	1.2	1.5
The minimum distance from the end of a loadbearing external wall to the edge of a door or window opening	1.0	1.0	1.2	1.5
The minimum distance from the end of a nonbearing external wall to the edge of a door or window opening	1.0	1.0	1.0	1.0
The minimum distance from the exposed corner of an internal wall to the edge of a door or window opening	1.0	1.0	1.5	2.0
The maximum height of a non-anchoring parapet wall (non-exit and non-entrance locations)	0.5	0.5	0.5	0.0

6.6 Construction Measures

6.6.1 Construction Measures for Multi-Storey Brick Masonry Buildings

Cast-in-situ reinforced concrete construction columns (hereinafter referred to as construction columns) should be installed in accordance with the following requirements for multi-storey masonry buildings:

1) Generally, construction columns should be installed in accordance with the requirements

listed in Table 6-8.

2) For buildings with few transverse walls, such as school buildings and hospitals, the construction columns should be arranged based on the number of storeys (Table 6-8).

The construction columns of multi-storey clay brick buildings should meet the following requirements:

Constructional column installation requirements Table 6-8

Number of storeys / Intensity				Installation location	
VI(6)	VII(7)	VIII(8)	IX(9)		
4 and 5	3 and 4	2 and 3		Four corners at the staircase and elevator room; bottom and top of the inclined ladder of a staircase; four corners of the external walls and corresponding corners; joints of the transverse walls and outer longitudinal walls at split-storey locations; joints of the internal and external walls in large rooms; both sides of larger openings	Every 12 m or joints between transverse walls and outer longitudinal walls; joints between inner transverse walls and outer longitudinal walls at the corresponding opposite side of a staircase
6	5	4	2		Joints between transverse walls (axes) and external walls in separate rooms; joints between gable walls and inner longitudinal walls
7	≥6	≥5	≥3		Joints between internal walls (axes) and external walls; partial smaller piers of internal walls; joints between inner longitudinal walls and transverse walls (axes)

1) For construction columns, the minimum cross-section should be 240mm×180 mm, a longitudinal reinforcement of 4φ12 should be used, the stirrup spacing should not exceed 250 mm, and more stirrups should be added at the top and bottom of the column if needed. For buildings exceeding six storeys with intensities 6 and 7, exceeding five storeys with intensity 8 and buildings with intensity 9, the longitudinal reinforcements of the constructional column should adopt 4φ14, the stirrup spacing should not be greater than 200 mm, and the section and reinforcement of the construction columns at the four corners of the building should be properly increased.

2) The joints between construction columns and walls should be constructed in a zigzag format, 2φ6 tie bars should be placed every 500 mm in the vertical, and the length of the tie bar extending into the walls at each side should be at least 1 m.

3) The construction columns should be connected to the ring beam, and the longitudinal reinforcement of the construction column should be continuous throughout the ring beam.

4) Individual foundations may not be installed for construction columns; however, the columns should extend 500 mm below the outdoor ground level or be anchored in the foundation ring beam located not more than 500 mm below the ground level.

For structures with a bay longer than 7.2 m with intensity 7 and all buildings with intensities 8 and 9, 2φ6 tie bars should be placed every 500 mm in the vertical and extended to the walls on each side with a length of at least 1 m at the corners of the exterior walls and the intersections of the interior and exterior walls.

For partition walls, 2φ6 bars should be used to tie the bearing walls, or columns should be arranged every 500 mm in the vertical and extend to the wall on both sides with a length of at least 500 mm. The top of the partition walls with longer than 5 m should be tied to the slab or the beam with intensity 8 or 9.

The cast-in-situ reinforced concrete ring beams in multi-storey masonry buildings should be installed in accordance with the following requirements (Fig. 6-9 and Fig. 6-10):

1) For buildings with precast reinforced concrete slabs and/or wooden roofs, the ring beams should be installed in accordance with the requirements set forth in Table 6-9 if the transverse walls are used for bearing. The ring beams should be installed in each storey if the longitudinal walls are used for bearing, and the spacing of the ring beams on the seismic transverse walls should be reduced accordingly, as listed in Table 6-9.

2) For buildings with cast-in-place RC slabs and roofs with reliable connections, ring beams are not necessary. However, the slab should be well connected to the corresponding construction columns.

3) For buildings with masonry arch slabs with intensities 6 and 7, the ring beams should be installed at each masonry wall for each storey.

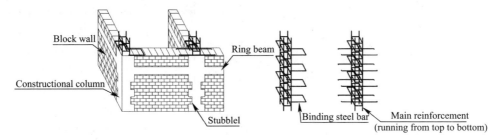

Fig. 6-9 Construction columns and ring beams

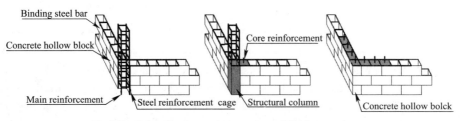

Fig. 6-10 Structural measures for concrete small block core columns

Requirements for installing cast-in-situ reinforced concrete ring beams in masonry buildings

Table 6-9

Type of wall	Intensity		
	VI(6) and VII(7)	VIII(8)	IX(9)
Exterior and interior longitudinal walls	At the roof level and floor of every storey	At the roof level and floor of every storey	At the roof level and floor of every storey
Interior transverse walls	As above; the spacing of the beam at the roof should not exceed 4.5 m; the spacing should not exceed 7.2 m at each floor; should be installed at the position of the construction column	As above; should be installed on the transverse wall of each storey and the spacing should not exceed 4.5 m; should be installed at the position of the construction column	As above; on the transverse walls of each storey

The construction of cast-in-situ reinforced concrete ring beams in multi-storey masonry buildings with clay bricks should comply with the following requirements:

1) The ring beams should be closed and need to overlap if there are gaps. The ring beams should be installed at the same level as the precast slab or close to the bottom of the slab.

2) If no transverse wall exists in the spacing requirement in Table 6-9, then the reinforcement for the beam or slab joint should be used in the replacement of the ring beams.

3) The cross-section height of a ring beam should be at least 120 mm, and the reinforcement should comply with the requirements in Table 6-10. The cross-section height for the foundation ring beam should be at least 180 mm, and the reinforcement should be at least $4\phi12$. For ring beams in buildings with brick arch floors and roof, the cross-section height and reinforcement should be determined by calculation, and the reinforcement should exceed $4\phi10$.

Requirements for the reinforcement of ring beams   Table 6-10

Reinforcement	Intensity		
	VI(6) and VII(7)	VIII(8)	IX(9)
Minimum longitudinal reinforcement	$4\phi10$	$4\phi12$	$4\phi14$
Maximum stirrup spacing (mm)	250	200	150

The floor and roof slabs in multi-storey masonry buildings should follow the requirements listed below:

1) Cast-in-situ reinforced concrete floors or roof slabs should extend to the longitudinal or transverse walls for a length of at least 120 mm.

2) Precast reinforced concrete floors or roof slabs should extend to the exterior wall with a length of at least 120 mm if the ring beam is not located at the same level as the slab. They should extend to the interior wall for a length of at least 100 mm.

3) If the span of a precast slab in the direction parallel to the exterior wall is larger than 4.8 m, then the side of the slab adjacent to the exterior wall should be tied to the exterior wall or the ring beam.

4) For large span slabs at the corners of a building, the precast reinforced concrete slabs should be tied to each other and tied to the beam, the wall or the ring beam if the ring beam is located at the bottom of the slab.

The reinforced concrete beams or trusses in the floors and roof system should be securely connected to the walls, columns (including construction columns) or ring beams. The connection between the beams and the brick columns should not weaken the column cross-section.

Sloping roof trusses should be securely connected to the ring beam at the top storey. Purlins or roof slabs should be reliably connected to the walls or trusses. Eave tiles above the entrance or exit of a building should be anchored with roof members. For intensities 8 and 9, on top of the longitudinal wall at the top storey, stepwise brick piers should be built to support the end gable wall.

Precast balconies should be reliably connected to the ring beam and the cast-in-place stripe in the floor slab.

Brick lintels should not be used at door and window openings, and the bearing length of the lintel should be at least 240 mm for intensities 6—8 and should exceed 360 mm for intensity 9. Staircases should be built in accordance with the following requirements:

1) A $2\phi6$ longitudinal reinforcement and tie meshes constructed by spot welding $\phi4$ distributed short reinforcement or $\phi4$ steel reinforcement mesh should be arranged every 500 mm in the vertical for staircase walls on the top storey. A 60-mm-thick reinforced concrete with a longitudinal steel reinforcement of at least $2\phi10$

or at least 3 layers of reinforced brick strips should be arranged at the stair landing or the half-height of the storey for the staircase wall with intensities 7—9. Each reinforcement strip should be of at least 2ϕ6, and the mortar strength grade should neither be less than M7.5 nor less than the mortar strength grade of the walls on the same storey.

2) The girder supporting length in the staircase and the exposed corners of hallway internal walls should be at least 500 mm, and the girder should be connected to the ring beam.

3) Fabricated staircases should be reliably connected to the beam of the landing slab and should not be used for intensities 8 and 9. Overhang in-wall stair tread or vertical fins for stair treads inserted into the wall should not be used.

4) In the staircase room and the elevator room above the roof of a building, the construction columns should be extended to the top and connected with the top ring beam. A 2ϕ6 reinforcement and a tie mesh constructed by spot welding a ϕ4 distributed short reinforcement or ϕ4 steel reinforcement mesh should be arranged every 500 mm in the vertical for the entire wall.

The same types of foundations should be used for the same construction units. Their bottoms should be buried at the same elevation. Otherwise, a foundation ring beam should be added and constructed with a slope of 1 : 2.

6.6.2 Constructional Measures for Multi-Storey Block Masonry Buildings

For small-sized concrete block buildings, the reinforced concrete core columns should be installed in accordance with the requirements listed in Table 6-11. For buildings with few transverse walls, in addition to the requirements in Table 6-11, the number of storeys should be increased by one or two.

Requirements for the installation of core columns in small-sized concrete block buildings

Table 6-11

Number of storeys Intensity				Arrangement position	Arrangement quantity
Ⅵ(6)	Ⅶ(7)	Ⅷ(8)	Ⅸ(9)		
4 and 5	3 and 4	2 and 3		Outer corners of external walls, four corners of buildings and elevator rooms, corresponding walls at the top and bottom of inclined ladders for staircases; Internal and external wall joints of a large-span room; Transverse and outer longitudinal wall joints at split-storey locations; Every 12 m or at unit transverse wall and outer longitudinal wall joints	Full grouting of 3 holes at the corners of external walls; Full grouting of 4 holes at the joints between internal and external walls; Full grouting of 2 holes at the corresponding walls at the top and bottom of inclined ladders for staircases
6	5	4		As above; Transverse walls (axis) and outer longitudinal wall joints in separate rooms	

Continued Table

Number of storeys Intensity				Arrangement position	Arrangement quantity
VI(6)	VII(7)	VIII(8)	IX(9)		
7	6	5	2	As above; Each internal wall (axis) and the outer longitudinal wall joints; Inner longitudinal wall and transverse wall (axis) joints as well as both sides of openings	Full grouting of 5 holes at the corners of external walls; Full grouting of 4 holes at the joints of internal and external walls; Full grouting of 4—5 holes at the joints of internal walls; Full grouting of 1 hole at both sides of openings
	7	≥6	≥3	As above; Inner core column spacing of transverse walls should not exceed 2 m	Full grouting of 7 holes at the corners of external walls; Full grouting of 5 holes at the joints of internal and external walls; Full grouting of 4—5 holes at the joints of internal walls; Full grouting of 1 hole at both sides of openings

Note: At the corners of the exterior walls, at the intersections of the exterior and interior walls or at the corners of the staircase or elevator shaft, construction columns can be substituted for core columns.

The core column in small-sized concrete block buildings should comply with the following construction requirements:

1) The core column section of small-sized concrete block buildings should be at least 120 mm× 120 mm.

2) The concrete strength grade of the core column should be of at least C20.

3) The vertical steel dowel of the core column should penetrate the wall and be connected to the ring beam. The dowel reinforcement should be of at least 1ϕ12, and it should be more than 1ϕ14 if it exceeds five storeys for intensities 6 and 7 or four storeys for intensity 8 or 9.

4) The core column should extend to a depth of 500 mm below the outdoor surface or be anchored to the foundation ring beam buried below the surface at most 500 mm.

5) The core columns installed for improving the seismic shear strength of a wall should be evenly distributed with spacing no greater than 2 m.

6) Steel meshes should be arranged at the wall joints of multi-storey small-sized concrete block buildings or the connections between core columns and walls. These meshes can be made by spot welding 4 mm-diameter steel bars. Their installation spacing in the vertical should not exceed 600 mm and should be arranged along the full horizontal length of the walls. For 1/3 storeys at the building bottom with intensities 6 and 7, 1/2 storeys at the building bottom with intensity 8 and all storeys with intensity 9, the spacing along the wall heights of the aforementioned

steel meshes should not exceed 400 mm.

In small-sized concrete block buildings, construction columns replaced with core columns should comply with the following requirements:

1) The construction column section should be at least 190 mm×190 mm. The longitudinal reinforcement should be of at least 4ϕ12 with a stirrup spacing no larger than 250 mm, and more stirrups should be added at the top and bottom of a column if necessary. For constructional columns in buildings more than 5 storeys high with intensities 6 and 7, more than 4 storeys high with intensity 8 and buildings with intensity 9, longitudinal reinforcements of 4ϕ14 should be used, and the stirrup spacing should not exceed 200 mm. The cross-section and reinforcement of the construction columns at the corner of the exterior walls should be properly increased.

2) The construction column and block wall joints should be built in a zigzag pattern. Masonry block holes adjacent to the construction columns should be filled for intensities 6 and 7. For intensities 8 and 9, steel bars should be added. Aϕ4 steel mesh reinforcement constructed using spot welding should be constructed every 600 mm in the vertical between the construction columns and the block walls and arranged along the full length of the walls. For 1/3 storeys at the building bottom with intensities 6 and 7, 1/2 storeys at the building bottom with intensity 8 and all storeys with intensity 9, the steel mesh reinforcement spacing in the vertical should not exceed 400 mm.

3) For joints between construction columns and ring beams, the longitudinal steel reinforcements of the columns should go through the longitudinal steel reinforcements of the ring beams.

4) Individual foundations should not be installed for each construction column; however, the columns should extend 500 mm below the outdoor ground level or be anchored in the foundation of the ring beam located at most 500 mm below the ground level.

6.6.3 Constructional Measures for Multi-Storey Masonry Buildings with Frame-Shear Walls at Lower Storeys

Reinforced concrete construction columns should be installed in the upper storeys of multi-storey masonry buildings with frame-shear walls at the lower storeys and comply with the following construction requirements:

1) The locations of the construction columns should follow the requirements listed in Table 6-8.

2) The cross-sections of the core columns should be at least 240 mm × 240 mm. The longitudinal reinforcement should be at least 4ϕ14, with a stirrup spacing no larger than 200 mm.

3) Construction columns should be installed in transition layers with reinforcements of at least 4ϕ16 for intensities 6 and 7 and of at least 4ϕ18 for intensity 8. Generally, the longitudinal reinforcement should be anchored in the lower frame columns, and if the anchor point is in the frame beam, then the relevant positions should be strengthened.

4) Construction columns should be reliably connected to the ring beams or the in-situ floor.

The floor slabs of buildings with frame-shear walls in the lower storeys should comply with the following requirements:

1) The floor slabs of the transition layer should be cast in situ with a width of at least 120 mm, and the number of openings should be small; furthermore if the dimension of the opening is larger than 800 mm, a boundary beam should be installed.

2) For other storeys, the cast-in-situ ring beams should be arranged when using fabricated reinforced concrete floor slabs; when cast-in-situ reinforced concrete floor slabs are used, an additional ring beam should remain unarranged. However, the reinforcement of the floor slab along the seismic wall periphery should be strengthened and stably connected to the corresponding construction columns.

Reinforced concrete breast summers in buildings with frame-shear walls at the lower storeys should comply with the following requirements:

1) The beam cross-section should have a width exceeding 300 mm and a height of at least 1/10 of the span.

2) The stirrup should be at least 8 mm in diameter with a maximum spacing of 200 mm.

3) In the vertical, lumbar should be laid at least $2\phi 4$ with a spacing not to exceed 200 mm in the beams.

If reinforced concrete walls are used at the bottom of masonry buildings with RC frames on the ground floors and seismic walls, then the section and construction should meet the following requirements:

1) The beams and frame columns should be laid along the circumference of the wall to form the frame. The frame beam should have a width of at least 1.5 times the thickness of the wall plate, and the height should be at least 2.5 times the thickness of the wall plate. Furthermore, the cross-section of the frame column should have a height of at least 2 times the thickness of the wall plate.

2) The wallboard thickness should be at least 160 mm or less than 1/20 of the wallboard clearance height; the wall should have set openings to form several wall sections, and the height-width ratio of each wall section should be at least 2.

3) The reinforcement ratio of the vertical and transverse distribution steel reinforcements for all walls should be at least 0.30% and adopt a double row arrangement; the spacing between the tie bars of the double row distributed steel reinforcements should not exceed 600 mm, and the diameter should be at least 6 mm.

If constrained brick masonry walls are used for the ground floor of intensity 6 brick buildings with frames on the ground floor and seismic walls, then the structure should be in accordance with the following requirements:

1) The thickness of the brick walls should be at least 240 mm, and the grade of the mortar should not be at least M10. The walls should be constructed earlier than the frame components.

2) A $2\phi 8$ horizontal reinforcement and tie meshes constructed by spot welding $\phi 4$ distributed short reinforcement should be arranged every 300 mm along the frame columns and the full horizontal length of the brick walls; the reinforced concrete horizontal tie beams connected to the frame columns should be arranged at the semi-height of the walls.

3) The reinforced concrete construction columns should be added inside the walls if the wall length exceeds 4 m and at both sides of the openings.

The enhanced grade of the material for buildings with lower frame-shear walls should comply with the following requirements:

1) The grade of concrete for the frame column and the seismic shear wall as well as the bressummer should be at least C30.

2) The strength grade of the masonry block materials in the transition storeys should be at least MU10; the mortar strength grade of the brick masonry should be at least M10, and the masonry mortar strength grade of block masonry should be at least Mb10.

[**Example 6-1**]

Seismic design example for a multi-storey masonry building A six-storey masonry building with a first-storey height of 3.85 m (from the foundation); the other storeys are 2.8 m high. The structural plan is illustrated in Fig. 6-11. The floors and roof are precast cored slabs. The thickness of the walls is 240 mm. The material of the upper storeys is multi-hole brick with a grade of MU10. The mortar grade is M10 for the first storey, M7.5 for the second and third storeys and M5 for the others. The dimensions of the doors and windows are illustrated in Fig. 6-11. The seismic intensity is 7 (the design of the basic earthquake acceleration is 0.10 g), and the earthquake is in Group 2 at a site of Category B. The dead load is 5.0 kN/m^2, and the live load is 2.0 kN/m^2. The representative values of the gravity loads of the storeys are $G_6 = 3345.2$ kN, $G_5 = 3785.9$ kN, $G_4 = 3785.9$ kN, $G_3 = 3785.9$ kN,

$G_2 = 3785.9$ kN, and $G_1 = 4245.2$ kN.

[SOLUTION]

1. Calculation of the horizontal seismic action

Based on the example earthquake, the horizontal seismic effect coefficient α_{max} is 0.08 according to the code. The total representative value of the gravity loads of the building is:

$$G_{eq} = 0.85 \sum_{i=1}^{6} G_i = 19323.9 \text{ kN}$$

The standard value of the total horizontal seismic action of the structure is:

$$F_{Ek} = \alpha_{max} \times G_{eq} = 0.08 \times 19323.9 = 1545.9 \text{ kN}$$

The standard values of the horizontal seismic action and the seismic shear force for each storey are indicated in Table 6-12 and Fig. 6-12.

2. Check for the seismic capacity of the walls

In the top storey, the wall section in axis ⑨ and between axis Ⓐ and axis Ⓔ must be checked. This can be denoted as wall 9. The area of wall 9 is as follows:

$$A_{top9} = (3.6 + 4.4 + 0.24) \times 0.24 = 1.98 \text{ m}^2$$

$$A_{top} = [(8 + 0.24) \times 3 + (3.2 + 0.24) \times 4 + (2 + 0.12) \times 4 + (4.5 + 0.24) \times 4 + (4.4 + 0.24) \times 4 + (6.8 + 0.24) \times 2] \times 0.24 = 23.65 \text{ m}^2$$

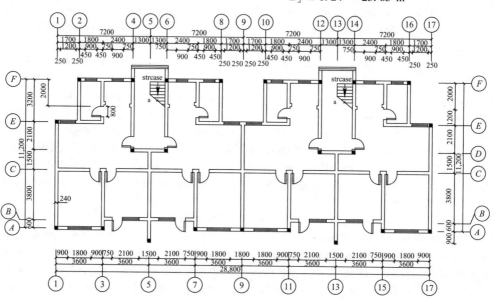

Fig. 6-11 Structural plan of the masonry building example (mm)

Standard values of the horizontal seismic action and the seismic shear force for each storey

Table 6-12

Item	G_i(kN)	H_i(m)	G_iH_i	F_i	V_i(kN)
6	3345.2	17.85	59,712	384.04	384.04
5	3785.9	15.05	56,978	366.45	750.49
4	3785.9	12.25	46,377	298.27	1048.76
3	3785.9	9.45	35,777	230.10	1278.86
2	3785.9	6.65	25,176	161.92	1440.78
1	4245.2	3.85	16,344	105.12	1545.9
Σ	22,734		240,364	1545.9	

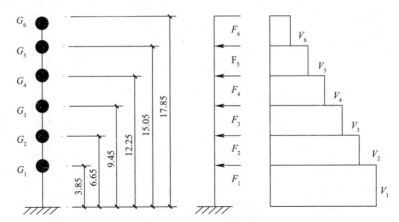

Fig. 6-12 Distribution of the seismic shear force (m)

The area of all of the transverse walls is 23.65 m². Furthermore, the architectural area of the top storey is as follows:
$$S = (11.2+0.24) \times (28.8+0.24) - 2 \times 1.7 \times 3.2 - (3.4-0.24) \times 3.2 - 4 \times 0.6 \times (3.6-0.24)$$
$$= 303.2 \text{ m}^2$$

The seismic load-bearing area of wall 9 is as follows:
$$S_9 = 3.6 \times (3.6+4.4+0.24) = 29.66 \text{ m}^2$$

$$\sigma_0 = \frac{8.24 \times 0.24 \times 1.4 \times 19 + (5 + 0.5 \times 2) \times 3.6 \times 8.24}{0.24 \times 8.24} = 0.117 \text{ N/mm}^2$$

From the code, the design value of the shear strength of the unit masonry M5 in a non-seismic design is $f_v = 0.11$ N/mm².
$$\sigma_0/f_v = 1.064$$

As listed in Table 6-1, the effect coefficient of the normal stress for masonry strength can be obtained as follows:
$$\xi_N = 0.998$$

The design value of the seismic shear strength along the stepped cross-section is as follows:
$$f_{vE} = 0.998 \times 0.11 = 0.11 \text{ N/mm}^2$$

Because the floor is semi-rigid, the seismic shear force of wall 9 can be calculated from the formula as follows:
$$V_{top9} = \frac{1}{2}\left(\frac{1.98}{23.65} + \frac{29.66}{303.2}\right) \times 384.04$$
$$= 34.9 \text{ kN}$$

The compression force of wall 9 at semi-height can be determined as follows (the volume weight of the brick masonry is 19 kN/m³):

For wall 9, the seismic adjusting coefficient for the load-bearing capacity γ_{RE} is assumed to be 1.0. Additionally, the partial coefficient for the horizontal seismic action is 1.3.
$$1.3 V_{top9} = 1.3 \times 34.9$$
$$= 45.36 \text{ kN} < \frac{f_{vE}A}{\gamma_{RE}}$$
$$= \frac{0.11 \times 0.24 \times 8.24 \times 10^3}{1.0}$$
$$= 217.5 \text{ kN}$$

Hence, the capacity of wall 9 is adequate.

Chapter 7
Seismic Design of Steel Building Structures

7.1 Introduction

In China, from 1949 to the 1980s, the application of steel structures was limited by steel production. However, with the increase of steel production, steel structures have been widely used in all types of structures (e.g., venues, stadiums, airport terminals, theaters, high-rise buildings, tower and mast structures, factory buildings, warehouses, houses, bridges and storage tanks) in the past 20 years due to the outstanding advantages of steel, such as low weight, high strength, good earthquake resistance, construction convenience and short construction time. In addition, the height of structures has been steadily increasing. For example, the 325-meter-high Diwang Tower in Shenzhen, the 421-meter-high Jin Mao Tower in Shanghai, the 492-meter-high World Financial Center in Shanghai and the 632-meter-high Shanghai Tower. Considering the diversity of steel structures, the steel structure buildings presented in this chapter refer to buildings whose primary load-bearing elements are steel elements. In this chapter, we use multi-storey buildings, tall buildings and single-storey steel factories as examples to introduce the seismic design principles of steel structures. Furthermore, there are certain buildings with steel roofs (roof trusses) that rest on reinforced concrete columns or brick walls and certain structures with reinforced concrete walls and steel columns that act together to resist horizontal forces; however, these structures are not considered in this chapter. The general principles of seismic design for a single-storey steel factory are basically the same as for factories with reinforced concrete columns. However, there are more sophisticated seismic design experiences for factories with reinforced concrete in China, and readers can refer to Chapter 8 for more details.

7.1.1 Earthquake Disaster Characteristics of Steel Structure Buildings

In general, the seismic performance of steel structure buildings is better than that of other buildings constructed with traditional materials because steel has good toughness and an excellent strength-weight ratio, which has been demonstrated to minimize damage in previous earthquakes. However, earthquakes can cause damage to steel structures if the design, material selection, construction or maintenance are not adequate, which is a lesson learned from previous earthquakes. For example, several steel structures were damaged during the 1994 Northridge earthquake in the United States. Also, the middle layer of several steel structures collapsed during the Han Shin earthquake in Japan. Based on the survey of earthquake damage of the Tangshan Steel factories after the 1976 Great Tangshan Earthquake in China, all-steel buildings with a total area of 36,700 m^2 did not collapse or were severely damaged; however, 9.3% suffered moderate damage (mainly due to buckling of struts and collapse of cladding walls). Among reinforced concrete structures with a total area of 40,600 m^2, 23.2% collapsed and were severely damaged, while 47.9% were moderately damaged. Among masonry structure buildings with a total area of 30,900 m^2, 41.2% collapsed or were severely damaged, while 20.9% were moderately damaged. The comparative data shows not only the good seismic performance of steel structures but also the importance of considering the potential earthquake damage to steel structures.

The earthquake damage to steel structures observed in previous large earthquakes can generally be divided into four types: structural member failure, joint failure, overall structure failure and non-structural member failure. Foundation failure, which is mentioned in the earthquake damage survey of steel structures, is introduced in Chapter 2 and is not discussed in this chapter.

1. Member Failure

In all previous earthquakes, most of the primary load-bearing structural members, such as beams, columns and braces, are partially damaged (Fig. 7-1), and the form of failure in the structural member includes buckling and fracture of braces, tensile failure and buckling of the joint panels, local buckling of flanges of beams and columns and horizontal cracking and brittle fractures of piers and plates. Fig. 7-1(a) illustrates the buckling of the roof cross-bracing using tensioned round steel under the action of repeated tension and compression, indicating the bending deformation of the members. Fig. 7-1 (b) depicts the damage of high-rise residential steel buildings in Hyogo-ken-Nanbu Earthquake in Japan in 1995: near the beam, column and bracing joints, fracture of the box-section column, and buckling of the ends of the H-shaped bracing. Fig. 7-1(c) illustrates the fracture of the lower chords of the trusses.

2. Joint Failure

Joint failure is one of the most common types of damage caused by earthquakes. The problems of stress concentration and strength imbalance are obvious because joints have a concentrated load transfer area and a complex design. In addition, welding and construction failures may occur; so, joint failure is very likely. Fig. 7-2(a) and Fig. 7-2(b) depict two different failure modes for a frame beam-column joint: the first is the failure of the weld bead connecting the beam and the column, while the second is the failure of the column stiffeners near the joint. The column holes in Fig. 7-2(c) appear because the connection joint plate of the bracing members and the web plate of the column are pulled apart. Fig. 7-2 (d) illustrates the flexural buckling of the gusset plate in the tubular truss.

Fig. 7-1 Member failure of steel structures

Fig. 7-2 Joint failure of steel structures

3. Overall Structure Failure

There are a few examples of the overall collapse of steel structures. Fig. 7-3 illustrates the deformation of the storey after a system failure, which is caused by plastic hinges and fracturing of the brace.

Fig. 7-3 Frame failure of a steel structure

4. Non-Structural Member Failure

A steel structure is never completely destroyed in large earthquakes because of its good load-bearing capacity and deformability, although the wall slabs, floor panels, roof panels, windows and doors attached to the structural members may be damaged. One reason for the damage of the non-structural members is that the strength of these members is not high enough or their deformability is not good. In addition, failure of the connection may occur.

7.1.2 Seismic Fortification Goal of Steel Structures

If a building is of great importance and not permitted to suffer any damage, the structure can be kept within an elastic range under the action of expected earthquakes. However, for the seismic design of most buildings, this requirement may not be practical, i. e., it is constrained by the economic situation and service conditions. An acceptable design objective under the current condition is that which takes full advantage of the good plastic properties of steel members and steel structure systems to ensure that the structure should keep its capacity to prevent collapse when subjected to expected major earthquakes.

A cantilever steel beam is subjected to an unidirectional concentrated force F (monotonic load) at the end of the beam, and its load-displacement δ curve is depicted in Fig. 7-4 (a). Point A on the curve is a mark point, which indicates that the steel beam reached its plasticity, and the steel beam reached its ultimate bearing capacity F_u at point B. Due to complete buckling or local buckling, the steel beam cannot bear any larger loads. The bearing capacity of the steel beam will gradually decrease if the beam displacement is forced to increase under laboratory conditions. It is apparent that the load F_y at point A is not the upper limit of the steel beam's bearing capacity. The curve for the steel beam will steadily rise as long as the load is less than F_u. The ratio between the displacement corresponding to the ultimate load of the steel beam and that corresponding to the yield load δ_u/δ_y is known as the ductility factor of the steel beam, which can reach 2—5 (or higher) for the members of steel structures.

If a statically indeterminate frame structure is subjected to an unidirectional horizontal distributed load F_i, the structure still holds a relatively high level of capacity to bear load (Fig. 7-4b) after several beams reach their upper bearing capacity limit (for example, after the beam forms plastic hinges at the end, the moment that the beam is subjected to will no longer increase) as long as the entire structure has enough redundancy. The ductility coefficient of frame structures can be defined similarly to the ductility coefficient of members. Clearly, as the ductility coefficient of the element or structure increases, the potential bearing capacity can be utilized after the elastic range becomes larger.

Earthquake motions are not monotonic. Experimental observations indicate that for random cyclic loading exceeding the elastic range of steel members or structures, the line segment

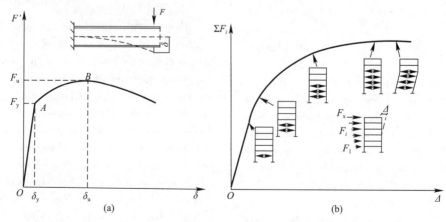

Fig. 7-4 Load-displacement curve of a steel structure subjected to a monotonic load
(a) Load-end displacement curve of steel beam; (b) Horizontal load-top displacement curve of frame

between the vertexes of the hysteresis curve generally coincides with the load-deformation curve under a monotonic load (Fig. 7-5). For most steel structures, the important task of the structural seismic design is to ensure that the structure, including its primary bearing members, remains within the elastic range or mostly within such a range under the action of frequent earthquakes; the force acting on the overall structure should not exceed its ultimate bearing capacity under expected rare earthquakes so that the structure can remain stable. To achieve this objective, we should consider the system layout, member design, joint construction and other factors.

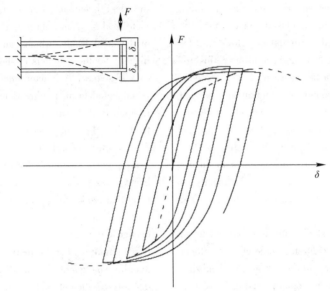

Fig. 7-5　Load-deformation curves of a steel beam under cyclic loading

7.2　Steel Structures for Middle and High-Rise Building

7.2.1　Seismic Conceptual Design

1. Structure System

There are several types of structures for multi-storey and high-rise steel buildings in a seismic design area, such as frame structures, braced frame structures, frame-seismic wall structures, frame-tube structures, tube-in-tube structures, bundled tube structures, truss-tube structures and mega structures (Fig. 7-6). Fig. 7-7 shows the typical plan sketches of some structures.

1) Moment-resisting frame structures

The moment-resisting frame structural system has the advantages of a simple configuration, a clear force transmission path, convenient construction and added flexibility in building layout as well as the window setup. In addition, it has considerable deformation and energy dissipation capacity under major earthquakes. The lateral stiffness of the frame system depends on the lateral stiffness of the frame columns and the flexural rigidity of the frame beams; hence, it is relatively smaller than other systems. The lateral deformation of low-rise structures is mainly shear deformation, which is mainly caused by the bending deformation of the frame

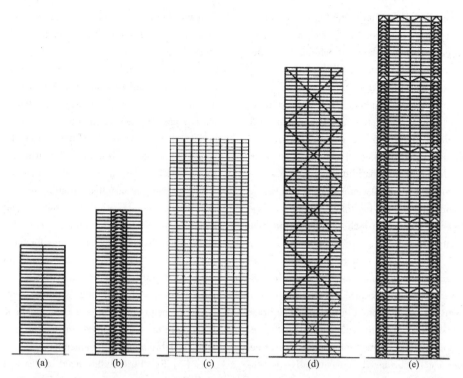

Fig. 7-6　Structure system of multi-storey and high-rise steel structure civil buildings: cross-section sketches
(a) Frame structure; (b) Braced frame structure; (c) Tube structure; (d) Truss-tube structure; (e) Mega structure

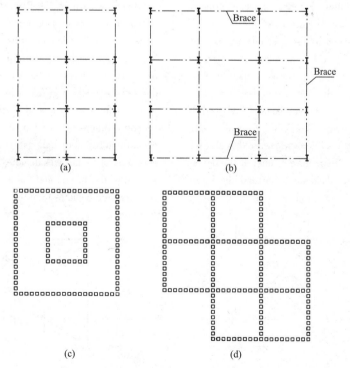

Fig. 7-7　Multi-storey and high-rise civil building structure systems: schematic plan
(a) Frame structure; (b) Braced frame structure; (c) Tube-in-tube structure; (d) Bundled tube structure

columns and the rotation of the joints. For high-rise buildings, the lateral deformation increases significantly because the axial deformation of the frame columns leads to bending of the overall structure, and the deformation belongs to the shear-bending deformation type. A suitable height for such frame structures is approximately 30 storeys, based on the comprehensive economic index and bearing capacity. The standard storey height of civil buildings is about 3.6 m; thus, the height of 30 storeys is about 110 m. The suitable height of the frame structure can be reduced in areas with high seismic fortification intensity.

2) Braced frame structures

Braced frame structures are formed by placing several evenly arranged braces along with the vertical and horizontal directions. The brace frame structure primarily relies on braces to resist horizontal earthquake motions. The steel braces can be arranged concentrically and eccentrically.

(1) Concentrically arranged braces

Fig. 7-8 shows sketches of concentrically braced frames, where the axes of the diagonals, columns and beams intersect at one point. The brace is typically made of steel and may be cross-shaped (Fig. 7-8a, or X-shaped), herringbone (Fig. 7-8b, or inverted-V-shaped), V-shaped (Fig. 7-8c), K-shaped (Fig. 7-8d), single oblique shape (Fig. 7-8e, f) and so on. Occasionally, it is difficult to ensure that all axes intersect in a real structure. Thus, if the deviation between the diagonal brace and the intersection of the beams and columns is no larger than the width of the brace member, then the brace can be considered as a concentrically arranged brace. A brace can only support axial forces under lateral load conditions. Furthermore, its axial stiffness and axial bearing capacity provide horizontal stiffness and horizontal bearing capacity to the structure. Normally, the horizontal stiffness of the braces exceeds the horizontal stiffness and horizontal bearing capacity provided by the columns of the frame. Thus, a braced frame is an effective structural system to resist horizontal earthquake motions, and the height of braced frame structures is considerably higher than that of frame structures for the same intensity zone. The beam-column connection of the braced frame structure is typically a rigid joint. However, for buildings with few storeys in a low-intensity zone, it is acceptable to use a hinge joint or a semi-rigid joint for the convenience of fabrication and connection.

(2) Eccentrically arranged braces

The hysteretic curve of a diagonal brace under cyclic loading after buckling due to compression is shown in Fig. 7-9. The bearing capacity under compression remains low, and the hysteretic loop demonstrates the so-called "pinch" phenomenon, where the curve is not as broad as the curves for beams and columns under a cyclic load (Fig. 7-5), indicating the poor plastic deformation capacity of the brace. Thus, eccentric braces are used in the seismic design of multi-storey and high-rise structures. In eccentric struts, the axis deviates from the intersection of the beams and columns (Fig. 7-10a, b), one side deviates from the intersection of the

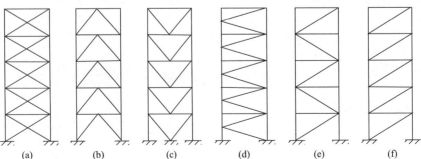

Fig. 7-8 Forms of concentrically arranged braces
(a) Cross-shaped; (b) Herring bone; (c) V-shaped; (d) K-shaped; (e) and; (f) Single oblique shape

beams and columns (Fig. 7-10c), or there is a gap between the brace and the original intersection for a herringbone brace (Fig. 7-10a). As indicated in Fig. 7-10(d), there is also another type of eccentric brace that extends a vertical segment downward from a steel beam and connects the brace to it. The member segments that are marked by the distance a in Fig. 7-10 are called the energy dissipation section or energy dissipation beam section, which has similar hysteretic characteristics as the beams and columns and can dissipate considerable amounts of plastic deformation energy.

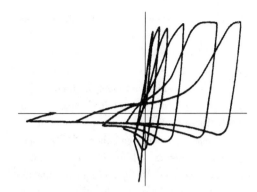

Fig. 7-9 Hysteretic curves for a brace member under cyclic tension and compression

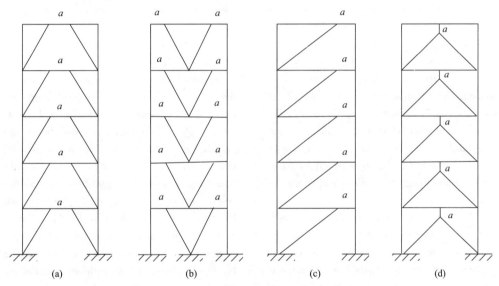

Fig. 7-10 Eccentric braces

Recently, engineering researchers have developed buckling-restrained braces, making it practical to overcome the decreased energy dissipation capacity resulting from the buckling of steel braces under compression.

(3) Frame-seismic wall structures

Setting seismic walls in frame structures has the same effect as braces. The seismic walls used in multi-storey and high-rise steel structures are medium panels or stiffened thin steel panels. Additionally, thin steel plate shear walls with vertical slits, reinforced concrete wall panels with vertical slits and reinforced concrete wall panels with embedded steel braces are used in seismic buildings. The shear walls in multi-storey and high-rise buildings are often placed in combination with staircases, elevator rooms and vertical fire escapes, with the majority being the reinforced concrete wall and the steel-reinforced concrete wall.

Reinforced concrete wall panels with vertical slits are constructed by setting vertical slits in the reinforced concrete wall panel at certain intervals and filling asbestos fibers as partitions in these vertical slits, which does not prevent vertical shear deformation but can act as noise insulation. The structure has a large lateral stiffness in the elastic stage, although it can yield and dissipate a large amount of energy during strong motions. Unlike solid shear walls, the stiffness degradation is small, and these structures have good ductility.

Reinforced concrete wall panels with embedded steel braces are comprised of prefabricated members, using steel plates as fundamental braces covered with a reinforced concrete wall panel. These members are only connected to the steel frame via joints, and there is a gap between the concrete wall panels and columns. The buckling inside the plane and outside the plane of the steel braces can be neglected due to the existence of a concrete cover, which can increase the bearing capacity as well as the lateral stiffness and dissipate the input energy of the earthquake.

(4) Tubular structures

Tubular structures have high stiffness and high lateral resistance. This system is an economical and effective structure because it can form a large functional space. Tubular structures can be divided into steel frame tubular structures, tube-in-tube structures, truss-tube structures and bundled tubular structures based on the arrangement, composition and number of tubes.

By placing columns at a distance of 3 m or less and using deep beams, the structure can resist lateral loads similar to that of a tube. The columns inside the tube only bear the force of gravity without providing resistance to horizontal loads. The overall structure has high stiffness and bearing capacity and can avoid the so-called "shear lag" phenomenon, in which the axial force in structures with sparse columns centralizes at the corner columns (Fig. 7-11). The tube-in-tube structure is a structure with double frame tubes (Fig. 7-7c). Bundled tube structures are composed of several tubes (Fig. 7-7d), which is an extension of the tube-in-tube structure concept. Truss-tube structures use large cross braces in the outer frame to transfer vertical shear forces and avoid the shear lag of frame tube columns, making it possible to decrease the distance between the columns and the height of the beams.

(5) Mega-frame structures

A mega-frame structural system is a novel system used in skyscrapers and it consists of three-dimensional truss columns with large column spacing and three-dimensional truss beams. The three-dimensional truss beams are arranged along the longitudinal and transversal directions to form a space truss layer. The sub-frame structure is arranged between the two space truss layers to support the load on each floor and transfer the load to the three-dimensional truss columns and beams through the columns of the sub-frame structure. The system has high stiffness and strength and can provide massive space in buildings.

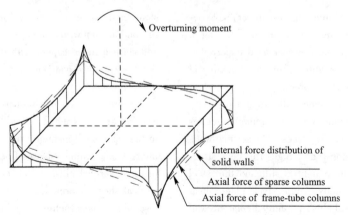

Fig. 7-11 Tube structure and "shear lag" phenomenon

2. Seismic Design Requirements

In addition to satisfying the general seismic design requirements for structures, the seismic design of structural steel structure buildings should meet the following requirements.

1) Maximum building heights

The maximum height of the different types of structural systems in different fortification intensity zones should conform to the regulations of the *Code for Seismic Design of Buildings* (GB 50011—2010) Version 2016 (Table 7-1). The building height in the table refers to the height from the ground level to the top plate of the primary roof panels (excluding the elevator room above the roof, antennas, etc.). The appropriate maximum height of steel structure buildings with irregular planes and vertical irregularities should be reduced accordingly; otherwise, the building must endure special reviews.

Maximum height of steel structure buildings (m) Table 7-1

Type of structures	Intensities VI(6) and VII(7) (0.10 g)	Intensity VII(7) (0.15 g)	Intensity VIII(8)		Intensity IX(9) (0.40 g)
			(0.20 g)	(0.30 g)	
Frame	110	90	90	70	50
Concentrically braced frame	220	200	180	150	120
Eccentrically braced frame (ductile plates)	240	220	200	180	160
Tube (frame-tube, tube-in-tube, truss-tube, bundled tube structure) and mega structure	300	280	260	240	180

2) Limitation to height-width ratios

The height-width ratio is the relationship between the total height of a building and the width along the direction of the horizontal earthquake motion. Since buildings may bear horizontal earthquake action in all directions, the height-width ratio is the ratio between the total height of a building and the smaller plan width. The concept of the total height of a building is consistent with the above height of the buildings. When the height-width ratio is large, the structural system is slender, and the lateral deformation under earthquake motions is large. The *Code for Seismic Design of Buildings* (GB 5011—2010) Version 2016 stipulates that the maximum height-width ratio of steel structure civil buildings should not exceed the values provided in Table 7-2.

3) Seismic grades

Steel structure buildings should use different seismic grades based on the fortification classification, intensity and height of the buildings in question and conform to the corresponding calculations and requirements. Buildings of Category C can be determined using Table 7-3.

Maximum height-width ratio Table 7-2

Intensity	Intensities VI(6) and VII(7)	Intensity VIII(8)	Intensity IX(9)
Maximum height-width radio	6.5	6.0	5.5

Seismic grade of steel structure buildings Table 7-3

Height (m)	Intensity			
	VI(6)	VII(7)	VIII(8)	IX(9)
≤50		4	3	2
>50	4	3	2	1

4) Setting seismic joints

In practical design, it is sometimes difficult to avoid irregularities of building frame system as expected. In this case, the seismic joint is necessary to prevent the disadvantages of the irregular layout. The gaps required for the expansion joint in steel structures should be at least 1.5 times larger than in RC structures.

5) Selection and configuration structural system

When selecting a structural system, multiple lines of defense must be set in the seismic design to resist earthquakes in addition to considering a suitable range for different heights. During the expected major earthquake, the damage to the structure can be accepted but the structure should keep stable even after the structural damage happens so that the gravity loads can be still supported by the structure.

In general, the development of plasticity and the formation of plastic hinges at the end of beams is expected to occur in frame structures because the beam failure mechanism, as indicated in Fig. 7-12(a), can dissipate considerably more energy than the column end plastic hinge failure mechanism, as indicated in Fig. 7-12(b), for multi-storey and high-rise frame systems. Beams should be considered as the first line of defense against earthquakes in a moment-resisting frame, therefore, the "strong-column-and-weak-beam type" is usually preferred. In the latter, the energy is simply dissipated in certain weak storeys of a structure. Consequently, the resistance mechanism of beams in frame structures is the first line to resist earthquakes.

In concentrically braced frame (CBF) structures (or frame-shear wall structures), because the lateral forces are mainly supported by braces, the braces provide most of the lateral stiffness. So, the bracing is the first resistant line. The eccentric braces dissipate the earthquake energy through the link, indicating that it is the first line of defense for eccentrically braced frame structures in the event of an earthquake.

In steel structure buildings with a braced frame structure, the lateral stiffness center should coincide with the center of the resultant horizontal seismic force. The braced frames should be symmetrical in two directions to reduce the potential torque. Furthermore, the length-width ratio of the floor between the braced frames should not exceed 3 to ensure that the lateral stiffness is uniformly distributed along the length direction and prevent in-plane deformations of the floors from affecting the lateral stiffness of the braced frame structure.

The steel structures in seismic Grade 1 or 2 should use braced frame structure or tube structure with eccentric braces, reinforced concrete seismic walls with vertical slits, reinforced concrete walls with embedded steel plates or buckling-restrained braces and other energy dissipation braces. Steel structures shorter than 50 m and in seismic Grade 3 or 4 should use concentrical braces, eccentric braces, buckling-restrained braces or other energy dissipation braces.

The concentrical braces of multi-storey and high-rise steel structures can be cross braces, herringbone braces or single oblique braces; however, K-shaped braces are not applicable due to their potential buckling under earthquake forces, which may cause large side sway and lead to the buckling or collapse. The axes of the concentrical braces, columns and beams should either intersect at one point or the deviation between the diagonal brace and the intersection of the beams and columns should not be larger than the width of the brace member. When single oblique braces, which can only bear tension, are used, two groups of diagonals in different slope directions should be used, and the difference in the projected area for each group of diagonals should be at most 10% to ensure that they have approximately the same resistance capacity under a lateral load. The buckling-restrained braces should use herringbone or single oblique braces in pairs, whereas the K-shaped or X-shaped braces are not applicable. Furthermore, the angle between the braces and columns should be between 35° and 55°.

A core structure surrounded by steel frames is also frequently used. In this case, the belt trusses and the linking trusses connecting the core and the surrounding frame are effective in making the frame resist overturning moment coupled with core structure (Fig. 7-13).

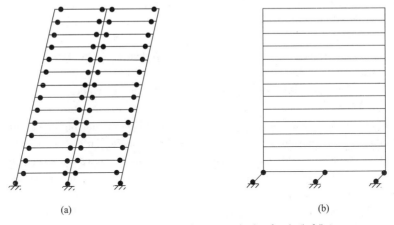

Fig. 7-12 Beam mechanism and storey mechanism for plastic failure

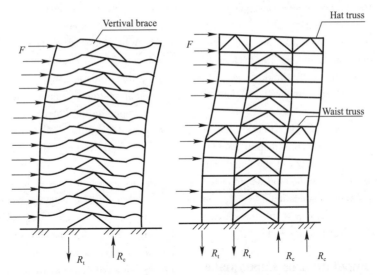

Fig. 7-13 Structure with strengthened storeys

6) Floor system selection

The seismic design of high-rise civil buildings requires a floor system with good integrity. Therefore, profiled steel sheeting cast-in-place concrete composite slabs or reinforced concrete slabs are before others because these slabs can provide sufficient stiffness at the floor plane which assures frames work as an entity. Profiled steel in non-composite slabs is only used as the bottom mold of the floor during construction, while the profiled steel of composite slabs can also replace the function of the steel bars at the bottom. In composite slabs, the profiled steel should have the ability to transfer the shearing force to work with the concrete slabs. These slabs must be reliably connected to the steel beams. A common technique is to implant studs on the steel beams. When profiled steel is used, these studs pierce through the steel plate to connect to the beams.

For steel structures shorter than 50 m, monolithic precast reinforced concrete and precast slabs or other lightweight slabs can be used at intensity 6 or 7. However, the embedded parts should be welded to the steel beams or other measures should be taken to ensure the integrity of a floor.

For conversion layer floors or slabs with large holes, the placement of horizontal braces, if necessary, is acceptable.

7) Setting the basement

When a basement is placed, the braces and seismic walls arranged vertically in the braced-frame (seismic wall plate) structure should extend to the foundation, and the frame columns should extend at least to the first floor below ground to ensure that the load can be transferred directly to the foundation. When the braces extend to the basement, the underground parts can use the braces covered by the reinforced concrete walls. When the frame columns extend downward, the underground parts can use the steel-reinforced concrete columns. In this case, the inner force can be transferred smoothly, which can ensure the interlocking of the pedestal and increase the rigidity, integrity and anti-overturning stability at the bottom of the building.

For steel structures higher than 50 m, a basement is required. When a natural foundation is used, the buried depth of the foundation should be at least 1/15 of the total height of the building; when a pile foundation is used, the buried depth of the pile cap should be at least 1/20 of the total height of the building.

7.2.2 Computation of Earthquake Action

After determining the layout of the structural system and preliminarily selecting the section size of the structure members, calculating the earthquake action primarily includes the following components: establishing the structural model for seismic calculations; selecting the ground motion parameters based on the requirements of the local earthquake resistance; selecting appropriate methods to calculate the earthquake action; determining the internal forces of the structural members and connections caused by the seismic effects and performing the seismic check calculation for the sections and connections; calculating and checking the deformation of the structure; and checking or determining the structural details based on the requirements put forth in the ***Code for Seismic Design of Buildings*** (GB 50011—2010) Version 2016.

1. Calculation Model

The plane structure model can be used for multi-storey and high-rise buildings if the structure is regularly arranged and its mass, as well as stiffness, are evenly distributed in the vertical, and the torque can be neglected. However, for structures with an irregular plane or elevation and a complex configuration as well as those that cannot be divided into planed lateral force-resistant units, the calculating model of spatial structures should be used. When calculating the structural effects, the bending deformation of beams and columns, the axial deformation of columns, the shear deformation of beams and columns and the influence on the lateral deformation due to the shear deformation of the panel zone should be considered. In general, the axial deformation of the beams is not considered unless the beams act as a belt truss or hat truss.

1) Simplified calculation model

The model for certain members in frame-bracing (seismic wall panel) structures can be simplified. The joints at the ends of oblique braces can be designed as rigid joints; however, for the calculations, these joints should be regarded as hinged joints for the small bending moment that they bear. Reinforced concrete panels with embedded steel braces are only connected to steel frames at the brace joints and can be simulated as braces. For reinforced concrete seismic panels with vertical slits, we can consider only the shear force generated by the horizontal load and neglect the compression generated by the vertical load.

2) Calculation of second-order effect

Steel structures can more easily produce large lateral deformations due to their flexibility compared to reinforced concrete structures. In this case, the product of the gravity load and the lateral displacement is called the gravity additional bending moment, which also denotes the so-called second-order effect (Fig. 7-14), and it can occupy a substantial part of the total moment. As a structure becomes taller, the corresponding second-order effect becomes larger. For any storey, if the second-order moment is greater than 1/10 of the overturning moment generated by the storey shear force timing the storey drift is taken into account, the second-order effect (P-Δ effect) should be considered in the calculation results.

3) Calculation of lateral displacement

The most commonly used section types for columns in high-rise steel structures are I-shaped (H-shaped), box (or rectangular) and occasionally circular tube. The areas of beam-column joints located in the column cross-section are known as the panel zones. Under a horizontal seismic action, the bending moment of the beam ends on both sides of a rigid joint are opposite and balance the bending moment of the column ends (Fig. 7-15a). A nodal bending moment causes a shear force in the panel zone, making the web of the column in the panel zone generate shear deformation (Fig. 7-15b).

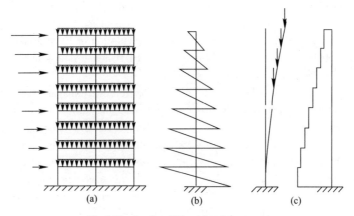

Fig. 7-14 Gravity additional bending moment
(a) Seismic behavior of high-rise steel structures; (b) Layer bending moments caused by horizontal forces;
(c) Additional layer bending moments caused by a gravity load

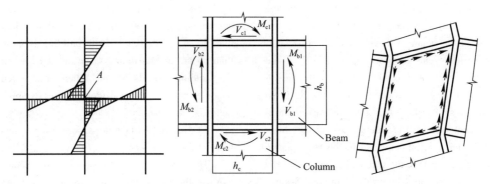

Fig. 7-15 Shear deformation of a joint-panel
(a) Bending moment and shearing force of joint-panel A; (b) Deformation of joint A

Due to the influence of this shear deformation, the calculated lateral displacement may decrease if we do not consider the deformation of the panel zone. The calculation and comparison indicate that if we take the distance between the axes as the member length in the elastic analysis, then the structural lateral displacement is overestimated due to the increasing rigidity of the connection. To a certain extent, it is canceled out by underestimating the frame lateral displacement caused by neglecting the shear deformation in the panel zone. For I-shaped section columns, the difference between considering shear deformation or not is as high as 10%. If we take the distance between the axes as the member length in the structural analysis for box section columns, the evaluation of the structural stiffness tends to be low. However, the shear deformation of the panel zone in box section columns is generally small, and its influence on a frame's lateral deformation can be neglected. Consequently, in the code, for I-shaped columns, the shear deformation of the panel zone should be considered. For box section column frames, concentrically braced frames and structures smaller than 50 m, the influence of the shear deformation in the panel zone does not need to be considered for the calculation of the inter-storey drift and can be approximately analyzed using the frame axes.

$$V_c = (M_{b1} + M_{b2})/h_b - (V_{c1} + V_{c2})/2$$
$$V_b = (M_{c1} + M_{c2})/h_b - (V_{b1} + V_{b2})/2$$

4) Influence of reinforced concrete floors on steel beam stiffness

When cast-in-place concrete floors are reliably connected to the steel beams, the interactions between the floors and beams should be considered in the elastic calculation. In the elastic analysis of steel frames, the moment inertia of the beam in the profiled sheeting composite slabs can be assumed as $1.5\ I_b$ if the beam is connected to the floors on both sides or $1.2\ I_b$ if the beam is connected to the floors on just one side, where I_b is the moment inertia of the steel beam. The interactions of the floors and beams can be neglected in the elastoplastic analysis because of the potential damage incurred at the connection between the floors and beams.

5) Horizontal forces of the frame in a braced frame structure

Based on the ***Code for Seismic Design of Buildings*** (GB 50011—2010) Version 2016, when verifying the cross-sectional strength of members against earthquake motions, we can consider only the results from the elastic analysis of frequently occurring earthquakes. In this case, the storey shear forces caused by horizontal seismic motions in braced frame structures are primarily undertaken by horizontal braces, which have higher stiffness. However, the braces are supposed to enter a nonlinear phase under rare earthquakes, and the horizontal forces undertaken by braces decrease while the proportion of horizontal forces undertaken by colums of the frame increases. Therefore, the design code requires that the lateral load carried by the frame is not less than the minimum of the following two values: 25% of the total base shear and 1.8 times the shear carried by the frame, calculated by elastic analysis.

2. Calculation of Earthquake Action

1) Damping ratio selection for steel structures

Several experiments have proven that structural damping varies not only with structural properties (e.g., damping is small when a structure is elastic, while damping tends to increase when a structure enters the plastic stage), but also with vibration severity; consequently, accurate determination of the structural damping is difficult.

By the experience, the seismic design code recommends that when the action under minor earthquake is to be computed, the damping ratio can be assumed to be 0.04 if the building is less than 50 m, while 0.03 if the building is more than 50 m but less than 200 m and 0.02 when the building exceeds 200 m. When an expected major earthquake is considered, the damping factor is taken as 0.05.

2) Calculation of internal forces for beams and column members

The moment at the design end of beams can use the moment at the edge of the column (Fig. 7-16), since the maximum moment occurs at this cross-section, which causes plastic deformation of the beam.

The design value of the inner forces for the link beam of the same frame beam in an eccentrically braced frame should use the product of

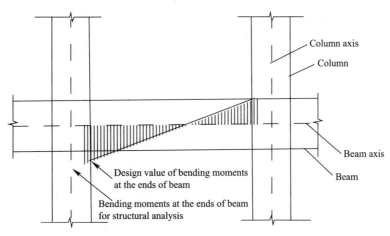

Fig. 7-16 Bending moments at the ends of a beam

the inner forces of the frame beam when the link beam reaches its shear capacity and amplification factor. The amplification factor should be not less than 1.3 for Grade 1, 1.2 for Grade 2 and 1.1 for Grade 3 to ensure that the structure only yields at the link beam. The shear capacity of the link beam is V_{tc} or V_c as calculated by Eq. (7-8). For example, if we assume that the design value of the bending moment is M_0 when we perform the seismic analysis and the shear force of the link beam is V, then the designed bending moments can be calculated as follows:

$$M = magnification\ factor \times \frac{V_{tc}(V_c)}{V} \times M_0 \quad (7-1)$$

where,

$\frac{V_{tc}}{V}$ or $\frac{V_c}{V}$ is at least 1.

The design value of internal forces for eccentrically braced frame columns should use the product of the internal forces of the frame column when the link beam reaches its shear capacity and the amplification factor. The amplification factor should not be less than 1.3 for Grade 1, 1.2 for Grade 2 and 1.1 for Grade 3, which can refer to Eq. (7-1) specifically. The seismic inner force of a steel frame column in a transference storey should be multiplied by a magnification factor of 1.5. All of these adjustments should prevent early damage to the beam segment adjacent to link in eccentric braced frame structures.

3) Inner force of brace members

The braced rods bear an axial force, considering the hinge joint of their edge. In a concentrically braced frame, if the eccentricity of the rod axis to the axis intersection of the beam-column does not surpass the width of the brace rods, it can still be considered as a concentrically braced frame; however, the additional bending moments should be considered.

The design axial force of eccentrically braced rods should use the product of the axial force of the eccentrically braced rods when the link beam connected with the brace reaches its shear capacity and an amplified factor. The amplified factor should be at least 1.4 for Grade 1, 1.3 for Grade 2 and 1.2 for Grade 3, which can refer to Eq. (7-1) specifically.

7.2.3 Seismic Checking for Members and Sections

1. Beams and Columns

In the equation for the cross-sectional strength and stability of the beams and column members, the left side is the combined value of the seismic effect and other effects, whereas the right side is the design value of the bearing capacity divided by the capacity coefficient for load-bearing γ_{RE}. Furthermore, when the strength calculations are performed, the value of γ_{RE} can be taken as 0.75; when the stability calculations are performed, the value of γ_{RE} can be taken as 0.80.

1) The bending capacity of beams and columns

To I- and H-shaped steel when bending around its strong section axis and box section:

When $N/N_y \leq 0.13$,
$$M_{pc} = M_p \qquad (7\text{-}2)$$
When $N/N_y > 0.13$,
$$M_{pc} = 1.15(1 - N/N_y)M_p \qquad (7\text{-}3)$$
When bending about its weak axis:
When $N/N_y \leq A_w/A$,
$$M_{pc} = M_p \qquad (7\text{-}4)$$
When $N/N_y > A_w/A$,
$$M_{pc} = \{(1-[(N-A_w f_{ay})/(N_y-A_w f_{ay})])\}M_p \qquad (7\text{-}5)$$
where,

N, N_y ——the design axial force and the axial yielding capacity, respectively; $N_y = A_n f_{ay}$;

A_n ——the net area of the section;

M_p, M_{pc} ——the full plastic moment without axial force and the ultimate moment with axial force existing, respectively;

A_w, A ——the web area and the total section area of the whole section, respectively;

f_{ay} ——the yield strength of the steel.

The designed bending moment M of the beams and column members should be at most the entire sectional bending bearing capacity M_{pc}, which can be confirmed by the above formulas. As indicated, when the axial force is zero on the members, we have $M_{pc} = M_p$.

2) Checking for strong-column-and-weak-beam conditions

To ensure that the full plastic bending capacity of the columns exceeds the full plastic bending capacity of the beam, and that buckling of the beams occurs before buckling of the columns, the full plastic bearing capacity at the ends of the beams and columns of the rigid joints should conform to Eq. (7-6) in addition to the following circumstances:

(1) The columns in a storey have the shear capacities 25% greater than those in the above storey;

(2) The design axial compression force does not exceed 0.4 times its design axial capacity, or the column, as an axially compressed member, can maintain its global stability under twice the design seismic action;

(3) The joints are connected with brace diagonals.

Beams with uniform sections:
$$\sum W_{pc}\left(f_{yc} - \frac{N}{A_c}\right) \geq \eta \sum W_{pb} f_{yb} \qquad (7\text{-}6a)$$

Beams with varied sections at the ends of the flange:
$$\sum W_{pc}\left(f_{yc} - \frac{N}{A_c}\right) \geq \sum (\eta W_{pb1} f_{yb} + V_{pb} s) \qquad (7\text{-}6b)$$

where,

W_{pb}, W_{pc} ——the plastic section modulus of the beams and columns, respectively;

W_{pb1} ——the plastic beam section modulus of the section where the plastic hinges are located;

N ——the axial compression design values of the columns;

A_c ——the cross-sectional area of the columns;

f_{yc}, f_{yb} ——the steel yield strength of the beams and columns, respectively;

η ——the column strengthening factor, which is 1.15 for seismic Grade 1, 1.10 for Grade 2 and 1.05 for Grade 3;

V_{pb} ——the plastic hinge shearing force of the beam;

s ——the distance between the plastic hinge and the surface of the column, and the location of the plastic hinge can use the minimum variable cross-section of the flange at the end of the beam.

2. Members of Centrically Braced Frames

(1) The compression bearing capacity of braced rods in concentrically braced frames can be checked as follows:
$$N/(\varphi A_{br}) \leq \psi f/\gamma_{RE} \qquad (7\text{-}7a)$$
$$\psi = 1/(1 + 0.35\lambda_n) \qquad (7\text{-}7b)$$
$$\lambda_n = (\lambda/\pi \sqrt{f_{ay/E}}) \qquad (7\text{-}7c)$$

where,

N ——the design axial force in the brace;

A_{br}——the area of the brace;

φ——the stability coefficient of the axially compressed steel members;

ψ——the reduced factor for strength considering the influence of cyclic load;

λ, λ_n——the slenderness ratio and regularization slenderness of the brace, respectively;

E——the elastic modulus of the brace;

f, f_{ay}——the design strength and the yield strength of the brace, respectively;

γ_{RE}——the adjusting coefficient for load-bearing of the brace, taken as 0.80.

(2) After the diagonals of the herringbone braces and V-shaped braces buckle by compression, the bearing capacity decreases sharply, and an unbalanced force will then be induced at the joint of the beams and braces. The unbalanced force will result in floor sag for herringbone braces and floor uplift for V-shaped braces. The herringbone or V-shaped brace crossbeam connections should remain continuous at the joint of the braces without considering the supporting function of the fulcrum when checking the bearing capacity under a lopsided force caused by buckling under compression and gravity. Furthermore, the value of the lopsided force should be 0.3 times the minimum yield capacity of the tensile braces and the maximum yield capacity of the compressive braces. The herringbone or V-shaped brace can be alternately set in the vertical, or zipper columns can be used if necessary.

3. Energy Dissipation Beams

The energy dissipation beams are critical for resisting an earthquake. After the buckling of the energy dissipation beams, other parts of a structure remain in the elastic condition. The energy from the dissipation beam breaking can dissipate the earthquake energy. The energy dissipation beams, which are short and have a high depth-span ratio, bear a large shear force, and the shearing capacity of the link beam can be checked using Eq. (7-8). The formula considers the situation in which the shearing capacity decreases based on the relevant relationships for the plastic section when there is an axial force applied at the end of the beam:

When $N \leq 0.15Af$,
$$V \leq \varphi V_t / \gamma_{RE} \quad (7\text{-}8a)$$
$$V_t = \min\{0.58 A_w f_{ay}; 2M_{tp}/a\}$$
$$A_w = (h - 2t_f) t_w$$
$$M_{tp} = W_p f$$

When $N > 0.15Af$,
$$V \leq \varphi V_{tc} / \gamma_{RE} \quad (7\text{-}8b)$$
$$V_{tc} = \min\{0.58 A_w f_{ay} \sqrt{1 - [N/(Af)]^2};$$
$$2.4 M_{tp}[1 - N/(Af)]/a\}$$

where,

V, N——the shearing force design value and the axial force design value of the link beam, respectively;

V_t, V_{tc}——the shearing capacity of the link beam and the shearing capacity considering the influence of the axial force, respectively;

M_{tp}——the full plastic bending capacity of the link beam;

a, h, t_w, t_f——the length, height of the section, thickness of the web, and thickness of the flange of the link beam, respectively;

A, A_w——the cross-sectional area of the link beam and the web, respectively;

W_p——the plastic section modulus of the link beam;

f, f_{ay}——the compressive strength of the steel design value and the yield strength of the link beam, respectively;

φ——the coefficient $\varphi = 0.9$;

γ_{RE}——the adjusting coefficient for load-bearing of the brace, $\gamma_{RE} = 0.75$.

4. Joints and Connections

1) Joint region

The yielding capacity of the panel zone should satisfy Eq. (7-9), which maintains an appropriate yield capacity. Using this aforementioned method, the panel zone will yield before the beam ends, and because the deformation of the zone will increase due to the small thickness of the web, adverse effects on the lateral displacements of the frame are avoided. Studies have indicated that a value of 0.7 times the value of the entire beam capacity is appropriate for the capacity of the panel zone. However, the value decreases to 0.6 times the value for seismic Grades 3 and 4, which avoids an unnecessary increase in the thickness of the web. Additionally, the relationship can reflect on the reduction coefficient ψ as follows:

$$\psi \frac{(M_{pb1} + M_{pb2})}{V_p} \leqslant \frac{4}{3} f_{yv} \quad (7\text{-}9a)$$

I-section column:
$$V_p = h_{b1} h_{c1} t_w \quad (7\text{-}9b)$$

Box-section column:
$$V_p = 1.8 h_{b1} h_{c1} t_w \quad (7\text{-}9c)$$

Tubular-section column:
$$V_p = \frac{\pi}{2} h_{b1} h_{c1} t_w \quad (7\text{-}9d)$$

The check of the panel zones for all three section columns should satisfy the following formulas:

$$t_w \geqslant (h_b + h_c)/90 \quad (7\text{-}10a)$$
$$(M_{b1} + M_{b2})/V_p \leqslant (4/3) f_v / \gamma_{RE} \quad (7\text{-}10b)$$

where,

M_{pb1}, M_{pb2}——the full plastic bending capacity of the beams on both sides of the panel zone;

V_p——the volume of the panel zone;

f_v——the design shearing strength of steel;

f_{yv}——the yielding shearing strength of steel, which takes 0.58 times the yielding strength of steel;

ψ——the reduction coefficient, which is 0.6 for Grades 3 and 4 and 0.7 for Grades 1 and 2;

h_{b1}, h_{c1}——the distance among the thickness midpoint of beam flanges and the thickness midpoints of column flanges (or the tube wall on the diameter line of steel tube), resprctively;

t_w——the thickness of the column web at the panel zone;

M_{b1}, M_{b2}——the design value of bending for the beam at the panel zone of both sides;

γ_{RE}——the seismic adjustment coefficient for the capacity of the panel zone, i.e., 0.75.

If Eq.(7-10) is not satisfied, then parts of the panel zone should be made of thick steel plates or reinforced by welded steel plates.

2) Rigid beam-to-column connections

The design of rigid beam-to-column connections should follow the principles of a strong connection for the weak components to guarantee the overall stability of a structure under earthquake motions. Specifically, the ultimate bearing capacity of all of the joints should be higher than the connected members. The calculation should be divided into two stages based on the ***Code for Seismic Design of Bwildings*** (GB 50011—2010) Version 2016.

(1) Elastic design

The design value of the bearing capacity of the member connections should be at least that of the connected members, and the connection with the high-strength bolts should not slip. Specifically, the connection and connected members should have equal strength. The calculation is performed based on the capacity of the members instead of the design forces in case of difficulties in the second phase of the connection design process, which result from the small design forces that may lead to small dimensions for the connecting pieces, few bolts or small effective cross-sectional sizes of the weld beads.

The check performed for the flanges and webs should be conducted separately, where flanges connected by a butt weld should satisfy Eq.(7-11), and webs connected by a fillet weld or bolts should satisfy Eq.(7-12), respectively:

$$\frac{M_F}{W_F} \leqslant \frac{f_t^w}{\gamma_{RE}} \quad (7\text{-}11)$$

$$\sqrt{\left(\frac{M_w}{W_w}\right)^2 + \left(\frac{V}{2 \times 0.7 A_w}\right)^2} \leqslant \frac{f_f^w}{\gamma_{RE}} \quad (7\text{-}12a)$$

$$\sqrt{\left(\frac{V}{n} + N_{My}\right)^2 + N_{Mx}^2} \leqslant N_v^b \quad (7\text{-}12b)$$

$$M_F = \frac{I_F M}{I_F + I_w}$$

$$M_w = \frac{I_w M}{I_F + I_w}$$

$$N_v^b = 0.9 n_f \mu p$$

where,

M, V——the moment design value and

the shear design value at the end of the beams, respectively;

M_F, M_w —— the moment design value of the flanges and the moment design value of the webs, respectively;

I_F, I_w —— the second moment of the beam flanges and the second moment of the beam webs, respectively;

W_F, W_w —— the resisting moment of the beam flanges and the resisting moment of the beam webs, respectively;

f_t^w, f_f^w —— the design value of the tensile strength of the butt weld and fillet weld, respectively;

A_w —— the net cross-sectional area of the beam webs;

n —— the number of high-strength bolts;

N_{My}, N_{Mx} —— the shear force induced by the design value of the bending moment undertaken by webs in the y and x directions for the most unfavorable bolt, respectively;

γ_{RE} —— the capacity adjustment coefficient of the attachment weld, $\gamma_{RE} = 0.75$;

N_v^b —— the shear capacity of each high-strength bolt;

n_f —— the number of friction surfaces that can transfer forces;

μ —— the anti-sliding coefficient of the friction surfaces;

P —— the design value of the pretension force for each high-strength bolt.

(2) Checking for yielding bearing capacity

The design capacity for the connections should exceed the design capacity of the connected members. Therefore, the first stage in the elastic design process of the rigid beam-to-column connections need to be checked using Eqs. (7-13) and (7-14) as follows:

$$M_u^j \geqslant \eta_j M_p \quad (7\text{-}13)$$
$$V_u^j \geqslant 1.2(2M_p/l_n) + V_{Gb} \quad (7\text{-}14)$$

The ultimate flexural capacity and the ultimate shearing capacity of the I sections can be calculated by using the approximation method which is used to calculate the flanges bearing bending moment and the webs bearing shear force. The ultimate flexural capacity M_u of full penetration groove welds on a beam's upper and lower flanges and the shear-resisting capacity V_u of the beam web connections can be calculated by using Eqs. (7-15), (7-16a) and (7-16b) as follows:

$$M_u = A_f(h - t_f)f_u \quad (7\text{-}15)$$

When fillet welds are used to connect the webs:

$$V_u = 0.58 A_f^w f_u \quad (7\text{-}16a)$$

When high-strength bolts are used to connect the webs:

$$V_u = nN_u^b \quad (7\text{-}16b)$$
$$N_u^b = \min(N_{vu}^b, N_{cu}^b)$$
$$N_{vu}^b = 0.58 n_f A_c^b f_u^b$$
$$N_{cu}^b = d\sum t f_{cu}^b$$

where,

M_u^j —— the flexural capacity of full penetration groove welds on a beam's upper and lower flanges;

V_u^j —— the shear-resisting capacity of the beam web connections;

M_p —— the plastic flexural capacity of the beams;

A_f^w —— the effective shear area of the fillet welds;

f_u —— the minimum tensile strength of the base material of the members;

A_f, t_f —— the sectional area and thickness of the flanges, respectively;

V_{Gb} —— the shearing force design value of the beam end section under the representative value of the gravity load (a high-rise building of intensity 9 should consider the characteristic value of the vertical earthquake action), which is defined by considering the beam to be a supported beam;

l_n —— the net span of the beam;

h —— the height of the beam;

N_u^b —— the shear capacity of each high-strength bolt at the connection nodes, which uses the smaller value of the shearing strength and the ultimate compressive strength of the bolt;

N_{vu}^b, N_{cu}^b —— the shear capacity of each high-strength bolt and the ultimate compressive strength of the corresponding connection panels;

η_j——the coefficient of the connection, which can be selected from Table 7-4;

n, n_f——the number of high-strength bolts and the number of shearing surfaces for the bolted connections, respectively;

f_u^b——the minimum tensile strength of the steel used for the bolts;

f_{cu}^b——the ultimate compressive strength of the connection panels, i.e., $1.5f_u$;

A_e^b——the effective cross-sectional area of the screw thread;

d——the bolt diameter;

Σt——the smaller value of the sum of the thickness of the steel plates along the same forced direction.

(3) Connections between bracings and frames

The connections between the bracings and frames should satisfy the following formulas, which provide the ultimate capacity of the splicing connection for columns, beams and bracings.

Connection and splice of bracings:
$$N_{ubr}^j \geq \eta_j A_{br} f_v \quad (7\text{-}17\text{a})$$
Connection of beams:
$$M_{ub,sp}^j \geq \eta_j M_p \quad (7\text{-}17\text{b})$$
Connection of columns:
$$M_{uc,sp}^j \geq \eta_j M_{pc} \quad (7\text{-}17\text{c})$$
The capacity of the connection between the foundation and the column bottom:
$$M_{u,base}^j \geq \eta_j M_{pc} \quad (7\text{-}18)$$
where,

M_p, M_{pc}——the plastic bending capacity of the beams and columns considering the effects of the axial load, respectively;

A_{br}——the area of the bracing section;

$N_{ubr}^j, M_{ub,sp}^j, M_{uc,sp}^j$——the ultimate compression (tension) and bending capacity for connection and splicing of the bracing, splicing of beam and column, respectively;

$M_{u,base}^j$——the ultimate bending capacity of the column bottom;

η_j——the connection coefficient, which can be obtained from Table 7-4.

Connection coefficients for the seismic design of steel structures   Table 7-4

Class of steel	Column and beam bracing		Connection of bracing and splice members		Column bottom	
	Welding	Bolting	Welding	Bolting		
Q235	1.40	1.45	1.25	1.30	Embedded	1.2
Q345	1.30	1.35	1.20	1.25	Wrapped	1.2
Q345GJ	1.25	1.30	1.15	1.20	Opened	1.1

If the yielding strength of the steel exceeds that of Q345 steel, then the coefficient should be obtained using the requirements for Q345 in Table 7-4. If the yield strength of the steel exceeds the value of Q345GJ steel, then the coefficients should be obtained using the requirements for Q345GJ in the table. If the connections are achieved using welding for flanges and bolting for webs, then the connecting coefficients should be obtained corresponding to the connection types in the Table 7-4.

7.2.4 Detailing Requirements for Seismic Design of Members

1. Beams and Columns

1) Slenderness ratios

The slenderness ratios for the frame columns of Grade 1 structures should not exceed $60\sqrt{235/f_{ay}}$, $80\sqrt{235/f_{ay}}$ for Grade 2, $100\sqrt{235/f_{ay}}$ for Grade 3, and $120\sqrt{235/f_{ay}}$ for Grade 4.

2) Width-thickness ratios for the plates of

beams and columns

The requirements in Table 7-5 should be satisfied for the plates of beams and columns in a frame. Table 7-5 is applied for Q235 steel; however, for steel of other classes, the values should be adjusted by multiplying by $\sqrt{235/f_{ay}}$. The formula $N_b/(Af)$ represent the axial force ratio of the beam.

Maximum width-thickness ratios for the plates of beams and columns in a frame

Table 7-5

	Plates	Grade 1	Grade 2	Grade 3	Grade 4
Column	Out flanges of I sections	10	11	12	13
	Plates of box sections	33	36	38	40
	Webs of I sections	43	45	48	52
Beam	Out flanges of I sections and box sections	9	9	10	11
	Flanges of box sections in the middle of two webs	30	30	32	36
	Webs of I sections and box sections	$72-120N_b/$ $(Af) \leq 60$	$72-100N_b/$ $(Af) \leq 65$	$80-110N_b/$ $(Af) \leq 70$	$85-120N_b/$ $(Af) \leq 75$

3) Lateral abutment

A lateral abutment should be formed to support the compression flanges of the beams and columns if necessary. Both the upper flanges and lower flanges should be supported at any cross-section of the members where plastic hinges might form. The out-of-plane slenderness ratios λ_y between the adjacent braced points should satisfy Eq. (7-19).

When $\quad -1 \leq \dfrac{M_1}{W_{PX}f} \leq 0.5$,

$$\lambda_y \leq \left(60 - 40\dfrac{M_1}{W_{PX}f}\right)\sqrt{\dfrac{235}{f_y}} \quad (7\text{-}19a)$$

When $\quad 0.5 \leq \dfrac{M_1}{W_{PX}f} \leq 1$,

$$\lambda_y \leq \left(45 - 10\dfrac{M_1}{W_{PX}f}\right)\sqrt{\dfrac{235}{f_y}} \quad (7\text{-}19b)$$

where,

λ_y —— the slenderness ratio under the action of out-of-plane bending, $\lambda_y = l_1/i_y$; where l_1 is the distance between the adjacent lateral support points; and i_y is the radius of gyration for the section;

M_1 —— the moment at the lateral support with a distance of l_1 from the plastic hinge: when the curvatures are in the same direction, $\dfrac{M_1}{W_{PX}f}$ is positive, and if the curvatures are in the opposite direction, then $\dfrac{M_1}{W_{PX}f}$ is negative.

When a frame is braced by herringbone or V-shaped braces, lateral bracing of the frame beam should be provided at the points connected to the brace. Furthermore, the slenderness ratios λ_y of the aforementioned points should satisfy the requirements above as well as the ratios of the supporting points between the beam ends.

2. Bracing Bars of Concentrically Braced Frames

1) Slenderness ratios

The slenderness ratios of the bracing bars should not exceed $120\sqrt{235/f_{ay}}$ when the bars are designed as struts, and the concentrical braces of Grades 1, 2 and 3 should not be designed as tie bars. If a brace of Grade 4 is designed as a tie bar, the slenderness ratio of the brace should not exceed 180.

2) Width-thickness ratios

The width-thickness ratios of the plates for the bracing bars should not exceed the values in Table 7-6. If the connection is provided by gusset plates, then the strength and stability of the gusset plate should be considered. The values in Table 7-6 are applied to Q235 steel. However, the values should be adjusted by multiplying by $\sqrt{235/f_{ay}}$ for other classes of steel and multiplying by $235/f_{ay}$ for tubular sections.

Maximum width-thickness ratios for plates of a concentrically braced steel structure   Table 7-6

Plates	Grade 1	Grade 2	Grade 3	Grade 4
Out flanges	8	9	10	13
Webs of I sections	25	26	27	33
Webs of box sections	18	20	25	30
Outside diameter-thickness ratios of tubular sections	38	40	40	42

3. Bracing Bars of Eccentrically Braced Frames

1) Slenderness ratios

The slenderness ratios of the bracing bars should not exceed $120\sqrt{235/f_{ay}}$

2) Width-thickness ratios

The width-thickness ratios for the plates of bracing bars should not exceed the limit of the axially loaded compression bars provided by the national standard ***Design Code for Steel Structures*** (GB 50017—2017).

4. Link Beams of an Eccentrically Braced Frame

Link beams are the primary members that dissipate energy, and their load-bearing capacity should be equal to that of the members connected by link beams. Furthermore, necessary measures are needed to retain their good hysteretic performance under cyclic loading conditions.

1) Steel with fine plastic yield should be used for the link beams instead of high strength steel with low elongation. Therefore, the yield strength for the steel of link beams should not exceed 345 MPa.

2) To consider the plastic property, the width-thickness ratios of the plates used for link beams and all other parts within the same span should not exceed the values in Table 7-7. The values in Table 7-7 apply to Q235 steel. However, for other classes of steel, the values should be adjusted by multiplying by $\sqrt{235/f_{ay}}$.

Maximum width-thickness ratios of the plates used for link beams and all other parts of the same span   Table 7-7

Plates		Maximum width-thickness ratios
Out flanges		8
Webs	If $N/(Af) \leq 0.14$	$90[1-1.65N/(Af)]$
	If $N/(Af) > 0.14$	$33[2.3-N/(Af)]$

3) The horizontal component of the axial force in the lateral bracing induces an axial force on the link beam. If the axial force is large, the length of the beam should be decreased to retain its good hysteretic performance in addition to decreasing the shear strength of the link beam using Eq. (7-8b). Thus, if the axial force of the link beam is $N > 0.16Af$, its length should satisfy Eq. (7-20).

When $\rho(A_w/A) < 0.3$,

$$a < 1.6M_{tp}/V_t \quad (7\text{-}20a)$$

When $\rho(A_w/A) \geq 0.3$,

$$a \leq [1.15-0.5\rho(A_w/A)]1.6M_{tp}/V_t \quad (7\text{-}20b)$$

where,

a——the length of the link beam;

ρ——the ratio of the design value of the axial load to the design value of shear force, $\rho = N/V$.

4) Link beams should not be strengthened

by welding steel plates, as the strengthened webs cannot enter the plastic stage.

5) Tapping is not permitted in link beams as it reduces the plastic deformation capacity.

6) At the connection of a link beam and brace diagonals, stiffeners should be set on both sides of the web, in which the length is the same as the height of the web (Fig. 7-17), to transfer the shear force and prevent buckling of the web of the brace diagonals under a concentrated force. The width of the stiffener should be at least $(b_f/2-t_w)$, and the thickness should be at least the larger value of $0.75t_w$ and 10 mm, where b_f is the width of the flange; and t_w is the thickness of the web in the link beam.

7) To prevent the link beams from buckling as well as local buckling induced by shearing and flexure, intermediate stiffeners can be used on the webs as follows:

(1) If $a \leqslant 1.6M_{tp}/V_t$, then the distance between the stiffeners should not exceed $(30t_w - h/5)$.

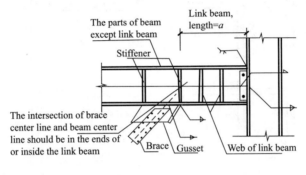

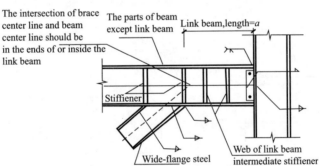

Fig. 7-17　Details of an eccentric brace and link beam

(2) If $2.6M_{tp}/V_t < a \leqslant 5M_{tp}/V_t$, then the intermediate stiffeners should be located $1.5b_f$ from the end of the link beams, and the distance should not exceed $(52t_w - h/5)$.

(3) If $1.6M_{tp}/V_t < a \leqslant 2.6M_{tp}/V_t$, then the distance between the intermediate stiffeners should be obtained through linear interpolation between the two values above.

(4) If $a > 5M_{tp}/V_t$, then the intermediate stiffeners may not be necessary.

(5) The length of the intermediate stiffeners should be the same as the height of the webs. If the height of the webs does not exceed 640 mm, the stiffeners should be set up on one side; otherwise, the stiffeners should be applied on both sides. The width of the stiffener should be at least $(b_f/2-t_w)$, and the thickness should be at least the larger value of t_w and 10 mm.

8) Both the upper and lower flanges of the ends of a link beam should be provided with out-plane bracings to prevent torsion. The design values of the axial load in the bracings should be at least 6% of the value of the product of the flange area and the design strength, i.e., $0.06b_f t_f f$. Furthermore, to retain the stability of the other parts within the same span,

the design values of the axial load should be at least 2% of the bearing capacity of the flanges when out-plane bracings exist, i. e., $0.02b_\mathrm{f}t_\mathrm{f}f$.

5. Details of Beam-Column Connections in Steel Frame Structures

Beam-column connections should use column-through joints or node parts to separate the columns, which is known as beam-through joints (Fig. 7-18). If the I-section columns are connected by rigid joints in one direction, the column webs should be placed in the plane of the frames with rigid joints.

The column-through rigid joints refer to Fig. 7-19, and the details can be given as follows:

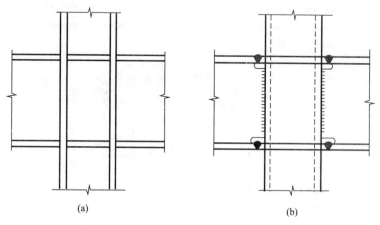

Fig. 7-18 Different forms of beam-column connections
(a) Column-through joint; (b) diaphragm-through joints

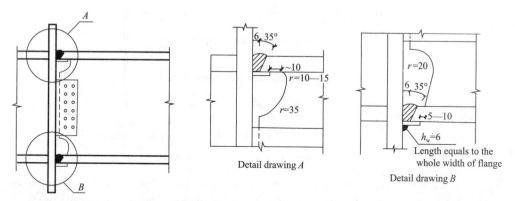

Fig. 7-19 Details of beam-column rigid joints for site connection

1) Structural members between beam flanges and column flanges should use a full-penetration V butt weld, for which the Charpy impact toughness should be at least 27 J at $-20^\circ\mathrm{C}$.

2) Transverse stiffeners should be set in columns in the corresponding position of the beam flange, and the thickness of the stiffeners should be at least that of the beam flange. When box-shaped sections are used and the stiffeners are inside the column, the stiffeners should be connected to the wallboards with a full-penetration butt weld. When an I-shaped section is used, the stiffeners should be connected to the column flanges with full-penetration butt welds and the webs with fillet welds.

3) The beam webs should connect to the column with high-strength bolts in slip-critical connections through the plate. The weld holes should be set at the corners of the webs, and the

holes should make the ends separate completely from the full-penetration butt welds between the flanges of the beams and columns.

4) When welding the webs between plates and columns, double filet welds should be used if the thickness of the plate is 16 mm or less, and the effective thickness of the welds should meet the strength requirements and be at least 5 mm. If the thickness of the plate is 16 mm or larger, K-butt welding should be used, which should use gas-shielded welding and end-profile return.

5) When the beams and columns are rigidly connected, the connecting weld between the column flange and web or the box-section column plate should be a full-penetration butt weld in the scope of 500 mm outside both the upper and lower beam flanges of the column.

6) For Grades 1 and 2, we should use flared connections, i.e., blind flanges, at the ends of the beam or bone-shaped connections, forcing the plastic hinge outside at the ends of the beam.

7) The distance between the splices of the frame columns and the top of the frame beams should use the minimum value of 1.3 m and half of the effective column height.

To ensure convenient installation on site, the rigid connection of beams and columns can also weld a cantilever beam to the columns. When the site installation of the columns is finished, it can be connected to the primary part of the beam at the ends of the cantilever beam. The connection can use all high-strength bolts in slip-critical connections or a combination of welding and bolts (Fig. 7-20).

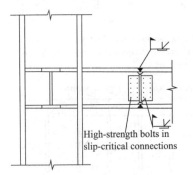
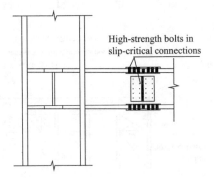

Fig. 7-20 Cantilever beam connection between frame beams and columns

7.3 Seismic Design of Steel Structures of Single-Storey Factories

7.3.1 Seismic Design Concepts

1. Structure System

Single-storey steel factories can be divided into horizontal structures (along the span direction) and longitudinal structures (along the column direction). For technological requirements, horizontal structures are column-beam systems or column-roof systems, whereas longitudinal structures are generally column-tie beam-brace systems. In modern large-scale industrial halls, a column-grid system is also used.

Horizontal structures can be divided into single-span and multi-span structures as well as shelving and frame systems. In the shelving system, a rigid connection is used at the base of the columns, whereas the roof connects to the columns using hinged connections. In the frame system, rigid and hinged connections can both be used at the base of the columns; the roof or solid web steel beams connect to the columns using rigid connections. When the solid web steel beams are used in multi-span structures, the connections can be partial rigid connections and partial hinged connections; however, at least two of the columns should be rigidly connected to the beams. The horizontal structure of a multi-span factory can be divided into equal height and unequal height as shown in Fig. 7-21.

Brace-hinged connection frames are frequently used in longitudinal structures (Fig. 7-22), in which the columns connect to the foundation using hinged connections, and the columns connect to the longitudinal beams, crane beams and brace members using hinged connections. Generally, brackets are used in factories with large column distances.

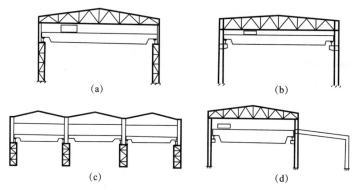

Fig. 7-21 Horizontal structure layout of a single-storey steel factory
(a) Single-span shelving structure; (b) Single-span frame structure; (c) Multi-span equal height plant;
(d) Multi-span unequal height plant

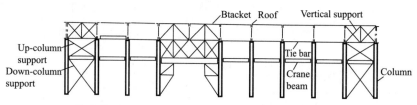

Fig. 7-22 Longitudinal structure layout of single-storey steel factory

Traditionally, roof systems of heavy industrial factories consist of a roof, braces and a reinforced concrete plate system, i.e., no-purlin system. For ventilation and lighting, a skylight may be needed. The rigidity, as well as the weight of such a structure, is extremely large. Currently, more factories use a fluted plate for the roof covering, which is light and retains heat. However, its rigidity is weaker than that of a standard roof system. It consists of a roof, braces, purlins and fluted plates, i.e., purlin system. If possible, transparent panels can be used to allow daylight to pass through, although the panels should be located in the same location as the profiled steel sheets, and ventilation problems can be solved using fans, which can eliminate the skylight structure, extend the roof and reduce vulnerable spots for the seismic design of factories.

Fluted plates are frequently used as walls in modern factories. However, certain factories still use blocks as walls. Even in factories with fluted plates, there are several structures using blocks above the foundation beam to 1—1.5 m. Additionally, the use of firewalls inside a factory built by blocks are not unusual.

2. Layout Requirements

With respect to seismic resistance, the following issues should be considered in the design of a structure:

1) Multi-span factories should be arranged with attention to height limitations whenever

possible. If height differences or construction (reduction of enclosure material) and labor (reduction of heating) costs can be reduced, a differential height can be used. However, the problems of low- and middle-stride roofs may apply a transverse force to the columns connecting the high and low strides under the influence of an inertia force in earthquakes, which should be considered in the design of steel columns.

2) The plane of interconnected factories should be regular. In the plane of the factory illustrated in Fig. 7-23, the longitudinal length is different for the two nearest spans, which may lead to transverse stiffness and mass of the framework in the 4-th and 5-th axes; the acceleration and displacement can be different, and the parts where the plane changes can be easily destroyed when subjected to seismic action in the direction of the span.

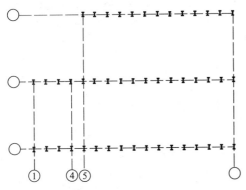

Fig. 7-23 Irregular factory layouts

3) Seismic joints should be used when the factory is complex. The width of the seismic joints should be 100—150 mm at the junction of the vertical and horizontal spans, as indicated in Fig. 7-24, for a factory without braces between columns. In other circumstances, the width can be 50—90 mm.

4) The crane ladder should not be near the seismic joints. The crane ladder of a multi-span factory should not be placed near the same horizontal axis. The ladder configuration determines the docked position. The axis where most cranes park will bear considerably more seismic forces than the others, which should be avoided in seismic design.

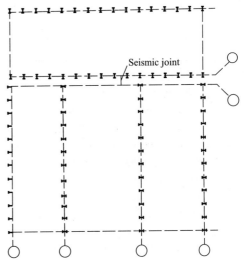

Fig. 7-24 Vertical and horizontal spans for perpendicular factory layouts

5) The lateral stiffness of columns should be uniform.

6) Horizontal braces should be used at the factory plane. Horizontal braces will not only reduce the calculation length of the out-of-plane steel beams or roof chords, but are also necessary to effectively transfer the horizontal seismic force into the steel columns.

If necessary, vertical braces should also be used to balance the roof height. They can also help transfer the horizontal seismic force into the columns.

In the following circumstances, longitudinal horizontal braces should be used:

(1) The spacing between the roofs exceeds 12 m;

(2) There are extra-heavy cranes or wall-lined cranes in the factory;

(3) There are heavy-duty cranes or intermediate-duty cranes in the factory;

(4) There is equipment that vibrates significantly in the factory;

(5) Large space stiffness is required;

(6) When there are brackets, the vertical bracing should be locally set at the brackets. Generally, longitudinal horizontal braces are set in the plane of the lower chord. There can only

be one on each side of the longitudinal direction for single-span factories, whereas there can be several on each side for multi-span factories.

7) In the plane of the longitudinal direction, bracings between the columns should be used if there are no other special exceptions. Because the length of the longitudinal direction is large and there may be heavy cranes, to reduce the impact of differential settlement, a reduction in the degree of indeterminacy should be attempted during the design process. Longitudinal members, such as cranes and connecting beams, should be linked to the columns with a hinge. The bending along the strong axis is generally in the plane of the transverse frame, whereas the secondary axis is in the longitudinal direction, in which the column-based is designed as a hinge. Thus, a bracing system is necessary to resist an earthquake in the longitudinal direction. A rigid frame is allowed to resist an earthquake only when the bracing system cannot be used, which can only be used in an area with intensity 6 or 7.

In factories with cranes, an upper and lower bracing system should be used between the columns in the middle of the longitudinal direction; furthermore, an upper bracing system should be used at the end of the longitudinal direction or where the lower bracing is located. In factories with lightweight roofs and walls, temperature stress is not usually a determining factor; however, the lower bracing can be used if necessary. When the length in the longitudinal direction exceeds 120 m (150 m when a lightweight envelop enclosure is used) in an intensity 7 or when the length is more than 90 m (120 m when a lightweight envelop enclosure is used) in an intensity 8 or 9 zones, at least one lower bracing should be positioned at 1/3 of the center.

7.3.2 Calculation of Earthquake Action

Generally, it is assumed that the horizontal frame of a structure bears the horizontal and vertical seismic forces, whereas the longitudinal frame bears the longitudinal seismic force.

1. Selection of the Structural Calculation Model

For most low-rise industrial structures, the horizontal component of an earthquake is considered in the main direction of ground motion, and the structure is usually simplified to the plane frame. Depending on whether a crane is present and whether the roof heights in the multi-span frame, a single-DOF model, two-DOF model, or multi-DOF model can be adopted.

Single-span or multi-span without a crane and an equal-height shelving system can be simplified into a single-particle cantilever column, as shown in Fig. 7-25. Additionally, other single-span or multi-span contour structures without cranes can be simplified as cantilever columns of a single particle; however, when determining the lateral stiffness of the equivalent columns, it is necessary to consider the effects of the beam rigidly connected to the column of interest, which is referred to as the restraint action.

Fig. 7-25 Single-particle model

Factories with cranes bear large amounts of gravity and horizontal seismic forces at the crane beam. In general, it can be considered as a dual-particle model as shown in Fig. 7-26. An unequal factory can be considered to be the same.

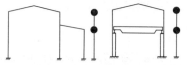

Fig. 7-26 Dual-particle model

In terms of structural stiffness, a single-storey steel factory can be simplified into a simple mechanical model. However, because of the difference in the column mass distributions, equivalent structural eigenvalues and equivalent inertial forces should be considered. Otherwise, the result will be unrealistic. Currently, computer programs can easily calculate structural eigenvalues, seismic forces and internal forces. Therefore, a multi-particle model can be constructed based on the actual situation. Except for the relatively simple situation mentioned before, as qualitative analysis and global master, simplifying the single-storey steel factory into single-particle or dual-particle models for the calculations is still meaningful.

2. The Consideration of the Weight and Stiffness of the Enclosure Material

If steel sheet or the precast concrete panels with flexible connectors are taken as the enclosure, only their weight shall be considered in the analytical model. In the case that the brick block is used and the necessary anchorage is embedded, the stiffness of the wall can be considered. Usually, the effective stiffness factor is assumed to be 0.6 for intensity 7.

3. Space Work Effective

The horizontal structures of a single-storey steel factory, which uses a pressure plate, can be regarded as independent structures. They are analyzed as either a shelving system or a frame. When large reinforced concrete roof plates are used, the rigidity of the roof should be considered. An equal-height factory can use the equivalent base shear method, whereas an unequal-height factory should use the modal response spectrum method.

4. Distribution of Earthquake Load in the Longitudinal Direction

For structures where lightweight walls or concrete panel with flexible connector is used as an enclosure, the following guidelines shall be followed for the distribution of earthquake load in another longitudinal frame:

(1) For reinforced concrete slab roof without purlins, the stiffness of roof structure should preferably be considered and the earthquake load is distributed as the stiffness of longitudinal frames.

(2) For roof material without enough stiffness, earthquake load is distributed according to the ratio of the mass that each longitudinal frame supports.

(3) For reinforced concrete slab roof with a purlin, the distribution method may adopt the average value of the results of the above two.

Factories with a common brick masonry enclosure wall against the columns will be discussed in Chapter 8. However, when a masonry wall is shorter than 1.5 m and the height of the remaining walls are the same as the lightweight walls, it can be considered as a lightweight wall factory.

5. Calculation of Vertical Earthquake Action

The *Code for Seismic Design of Buildings* (GB 50011—2010) Version 2016 stipulates that the vertical seismic force should be calculated when the roof span exceeds 24 m. Standard values for the vertical seismic force use the product of the representative value of gravity and the vertical seismic coefficient. The coefficients are specified in Table 4-12.

6. Crane Load Computing

The gravity force of the cranes and the hanging mass is added to the structure in terms of the inertial force caused by the earthquake. When calculating the combination of the seismic action, the representative gravity load includes the gravity force of the cranes and the hanging mass. The seismic action caused by gravity should be included in the seismic action calculations. For the hanging mass, based on the *Code for Seismic Design of Buildings* (GB 50011—2010) Version 2016, the hanging gravity of soft hook cranes can be ignored; however, the combination coefficient of gravity for hard hook cranes should be 0.3, although the actual value should be used if the mass is too heavy. In the static force calculation, there are at most two cranes that can be considered in one transverse frame calculation, and it is similar to the seismic calculation.

7. Damping Ratio

Based on the roof and enclosure wall types, the damping ratio values range from 0.045 to 0.050. The value should be small when a lightweight roof and enclosure walls are used.

7.3.3 Checking Members and Details

The indeterminate number of single factories is not very large; thus, the principle of member checking is similar to that of a determinate structure. However, under rare earthquakes, the members may enter the plastic stage, and the design of the members and junctions must follow the design code.

1. Limitation of the Slenderness Ratio for Steel Columns

If the slenderness ratio of a column is too large, its bearing capacity tends to decrease. Furthermore, it will obstruct the development of plastic deformation. The *Code for Seismic Design of Buildings* (GB 50011—2010) Version 2016 stipulates that the slenderness ratio of a column should not exceed 150 when the axial compression ratio is less than 0.2. If the axial compression ratio is more than 0.2, then the slenderness ratio of the column should not exceed $120\sqrt{235/f_{ay}}$, where f_{ay} is the yield stress of steel.

2. Limitation of the Width-Thickness Ratio for Beams and Columns

The plate width-thickness ratio directly affects the bearing capacity and deformability of steel bar members. Sections with a high plate width-thickness ratio, in which a large modulus of the section and moment of inertia can be obtained using extremely small section areas, are good for section flexural capacity and bar rigidity. However, elastic or plastic local buckling can potentially occur under a low inertial force, which will influence the capacity and the plastic utilization of the steel bar members. Therefore, when the structure can bear an applied force beyond the elastic stage, it is necessary to form a more stringent limitation for the plate width-thickness ratio. The *Code for Seismic Design of Buildings* (GB 50011—2010) Version 2016 ensures certain provisions are stricter than the elastic static design for the width-thickness ratio of a single-storey factory beam and column. The plate width-thickness ratio for factories with heavy roofs can be determined using Table 7-5: Grades 4, 3 and 2 of seismic measure can be taken in intensities 7, 8, and 9 zones, respectively. For factories with lightweight roofs, the plate width-thickness ratio limitation in the plastic energy consumption area can be determined by the performance objective based on the bearing capacity. The plate width-thickness ratio limitation out of the plastic energy consumption area can be obtained based on the plate width-thickness ratio limitation in the elastic design mentioned in the *Design Code for Steel Structures* (GB 50017—2017).

3. Checking and Details of Braces and Joints between Columns

The lower column bracing of factories with cranes should be made of structural steel, whereas the upper column bracing can be structural steel or round steel. The type of bracing should be crossed; however, V-shaped, inverted-V-shaped or other types of steel braces can also be used in a limited manner. The angle between the diagonal of the bracing and the horizontal plane should be less than 55°. When crossed lower column bracing cannot be used in heavy factories due to technological limitations, a gantry brace or chevron brace can also be used. The limitations of the slenderness of the cross braces are provided in Table 7-8.

Limitation of the slenderness of the cross braces   Table 7-8

Position	Seismic intensity			
	Intensities VI(6) and VII(7) of zones I and II	Intensity VII(7) of zones III and IV; Intensity VIII(8) of zones I and II	Intensity VIII(8) of zones III and IV; Intensity IX(9) of zones I and II	Intensity IX(9) of zones III and IV
Brace of upper column	250	250	200	150
Brace of lower column	200	150	120	120

Cross braces with a slenderness of less than 200 can be regarded as a tension member in an earthquake. Its tension design value N_t can be determined as follows:

$$N_t = \frac{l_i}{(1 + \psi_c \varphi_i) s_c} V_{bi} \quad (7\text{-}21)$$

where,

N_t——the axial tension design value of internode i;

l_i——the length of internode i;

ψ_c——the member uninstall coefficient;

φ_i——the axial compression stability factor of the cross brace in internode i;

V_{bi}——the design shear value of internode i;

s_c——the clear distance of the columns where the brace exists.

The inter-columnar bracing should be the entire structural steel. Welding with a constant strength, i. e., a butt weld, can be used if the steel oversteps the size limitation.

It is necessary to set a joint plate at the junction of the crossed column bracing, for which the thickness should be at least 10 mm. The link between the diagonal bar and the crossed joint plate as well as the link between the diagonal and the end joint plate should be welding. When there is a bar interruption in the crossed bracing, the crossed joint plate should be strengthened to bear a load that will be greater than 1.1 times the capacity of the bars.

The strength reduction factor should be considered if a single angle is used in the end link of the crossed bracing. For bars designed as tensile members, the strength reduction factor should be 0.85. It is forbidden to use a one-way eccentric connection in intensities 8 and 9 zones. The capacity of the link between the bracing and the elements should be at least 1.2 times the plastic capacity of the bracing bars.

The cross-sectional stress ratio of the braces should be at most 0.75.

4. Roof Braces and Details

In the cross-section design of the roof bracing, the crossed diagonal bars can be designed as tensile bars based on the internal force and using the non-coefficient area. In a roof with vertical bracing, the web members should bear the horizontal earthquake forces. The connectional capacity should be greater than the internal force of the bars, and the connection should meet the detailing requirements.

5. Column Feet Design

The column feet of a single-storey factory can be divided into submerged, outsourced and exposed, as indicated in Fig. 7-27. Submerged and outsourced column feet are rigid, whereas exposed feet can be divided into rigid and hinged column feet.

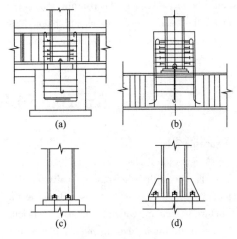

Fig. 7-27 Column feet
(a) Submerged; (b) Outsourced; (c) Exposed; (d) Exposed (rigid)

When the seismic intensity is 6 or 7, the column feet can be exposed. The column feet anchors should be multiplied by 1.2. No shrinkage mortar should be used for secondary grout between the bottom of the column and the top of the foundation. If the shear is too large, then a shear key should be used. Considering that the exposed column feet have been pushed out in past earthquakes, they should be used cautiously.

When solid web steel columns are inserted into the column feet, its embedded depth must be at least 2.5 times the height of the steel column or 0.5 times the width of the column. Furthermore, it should satisfy Eq. (7-22):

$$d \geq \sqrt{6M/b_f f_c} \quad (7\text{-}22)$$

where,

d——the embedded depth of the column feet;

M——the ultimate moment when the column feet yield across the section;

b_f——the bending flange width of the column;

f_c——the design value of the foundation concrete compressive strength.

6. Other Connections and Details

The *Code for Seismic Design of Buildings* (GB 50011—2010) Version 2016 specifies the requirements for other connections in a single-storey steel factory:

(1) The upper column joints with the frame should be located where the bending moment is small, the bearing capacity should be at least the inter force of the joints, which is calculated when the two ends of the upper column portions have a complete plastic cross-section, and it should be at least 0.5 times the tensile yielding bearing capacity when the columns yield across the entire cross-section.

(2) For the roof beam splice of a rigid frame, when the position is outside the maximum stress area of the beam, it is preferable to design it using the same strength as the spliced section.

(3) For rigid connections between solid web roof beams and columns and between the splices of beam ends and beams, it is preferable to design it in an elastic stage using the combination of the internal seismic force.

7.4 An Example of Seismic Design

[**Example 7-1**]

A four-storey steel frame is located at a Class Ⅲ site with a seismic intensity 7. The characteristic period is 0.55 s. Because the structure is regular, without a large eccentricity of mass distribution, the seismic calculation can be performed in two directions. The calculation for the transverse frame in Fig. 7-28 is required.

The elevation of the structure's indoor floor is ±0.00 m, and the elevation of the upper surface of the foundation is −0.5 m. The longitudinal frame intercolumniation is 8 m. An embedded column footing is used for the steel columns, and the steel columns are connected to the base by rigid joints. All of the beam-to-column connections are rigid and welded. The section dimensions and geometric features of the beams and columns are illustrated as follows:

The material of the steel beams is Q235B, and the material of the steel columns is Q345B.

The section dimensions of the members are as follows:

Z1: H500×400×12×16
Z2: H500×400×12×16
L1: H650×300×10×16

The dead loads and live loads of the structure can be given as follows:

Dead load:

The weight of each floor (including secondary beams, floors and surface weight and the suspension load): 5 kN/m²

Wall weight:
Exterior wall of axes 1 and 4   1.5 kN/m²
Interior wall of axes 2 and 3   1.0 kN/m²
Live load: FL4   1.5 kN/m²
            FL1-3   5 kN/m²

Members	$A(\text{cm}^2)$	$I_x(\text{cm}^4)$	$W_x(\text{cm}^3)$	Dead load(kN/m)
Z1	215.2	10,196.9	4077.9	1.689
Z2	184.16	85,239.5	3409.6	1.446
L1	157.8	116,159	3574.1	1.239

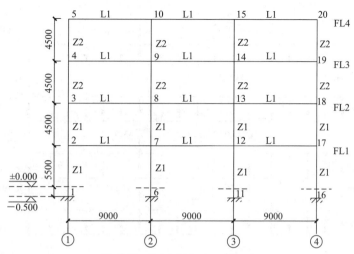

Fig. 7-28 Horizontal framework (mm)

[Solution]:
1. **Calculation of the deformation and internal forces**

1) Vertical load

The load of each floor:

Uniformly distributed load of floor FL1:

Dead load: 5.0×8+1.239×1.15=41.42 kN/m (1.15 is the increasing coefficient considering the structural parts of the steel beams).

Live load: 5.0×8=40 kN/m.

Concentrated load of columns:

Axes 1 and 4: 1.5×8×(2.25+2.5)+1.689×1.10×(2.25+2.75)=66.29 kN (1.10 is the increasing coefficient considering the structural parts of the steel columns).

Axes 2 and 3: 1.0×8×(2.25+2.5)+1.689×1.10×(2.25+2.75)=47.29 kN.

The uniformly distributed load of floor FL2 is the same as FL1.

Concentrated load of columns:

Axes 1 and 4: 1.5×8×4.5+(1.689×1.486)×1.10×2.25=61.76 kN.

Axes 2 and 3: 1.0×8×4.5+(1.689×1.486)×1.10×2.25=43.76 kN.

The uniformly distributed load of floor FL3 is the same as FL1.

Concentrated load of columns:

Axes 1 and 4: 1.5×8×4.5+1.446×1.10×4.5=61.16 kN.

Axes 2 and 3: 1.0×8×4.5+1.446×1.10×2.25=43.16 kN.

Uniformly distributed load of floor FL4:

Dead load: 5.0×8+1.239×1.15=41.42 kN/m.

Live load: 1.5×8=12 kN/m.

When calculating the gravity load, the live load of the roof should not be considered; however, the basic snow pressure of this area is 0.2 kN/m², and the live load can be replaced by 0.5×0.2×8=0.8 kN/m² (the parameter 0.5 is the combined factor that accounts for the snow load). The concentrated load of the columns is half that of FL3. Based on the above calculation, the distribution of the gravity load used for the seismic check of the sections is illustrated in Fig. 7-29, in which the live load is multiplied by 0.5 for reduction.

2) Seismic effect calculation for each floor

The gravity distribution of each floor is depicted in Fig. 7-30.

G_1 = 66.29×2+47.29×2+61.42×27 = 1885.50 kN

G_2 = 61.76×2+43.76×2+61.42×27 = 1869.38 kN

G_3 = 61.16×2+43.16×2+61.42×27 = 1866.98 kN

G_4 = 30.58×2+21.58×2+42.22×27 = 1244.14 kN

G_{eq} = 0.85 $\sum G_i$
   = 0.85×(1885.50+1869.38+1866.98+1244.14)
   = 5836.1 kN

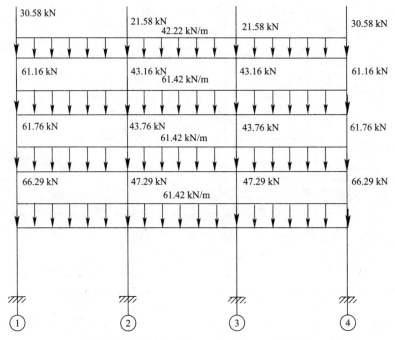

Fig. 7-29 Distribution of the representative gravity loads

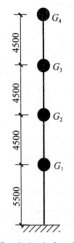

Fig. 7-30 Gravity load of each floor (mm)

Using finite element analysis, we can determine that the natural vibration period of the structure is $T_1 = 1.19$ s. Because the height of the structure is less than 40 m, its mass and stiffness are evenly distributed, and the deformation is primarily shear. The equivalent base shear method can be used for the seismic calculation of each floor. The characteristic period is 0.55 s, $T_g < T_1 < 5T_g$, and the seismic influence coefficient is as follows:

$$\alpha_1 = \left(\frac{T_g}{T_1}\right)^\gamma \eta_2 \alpha_{max}$$

where $T_1 = 1.19$ s and the damping ratio is 0.04 based on the *Code for Seismic Design of Buildings*(GB 50011—2010) Version 2016.

We can obtain the following:

$$\gamma = 0.9 + \frac{0.05-\xi}{0.3+6\xi} = 0.9 + \frac{0.05-0.04}{0.3+6\times0.04}$$
$$= 0.919$$

$$\eta_2 = 1 + \frac{0.05-\xi}{0.08+1.6\xi} = 1 + \frac{0.05-0.04}{0.08+1.6\times0.04}$$
$$= 1.069$$

$$\alpha_1 = \left(\frac{T_g}{T_1}\right)^\gamma \eta_2 \alpha_{max}$$
$$= \left(\frac{0.55}{1.19}\right)^{0.919} \times 1.069 \times 0.08 = 0.042$$

$F_{Ek} = \alpha_1 G_{eq} = 0.042 \times 5836.1 = 245.12$ kN

Because $T_1 = 1.19$ s $> 1.4T_g = 1.4 \times 0.55 = 0.77$ s, the top additional coefficient is given by:

$\delta_n = 0.88T_1 + 0.01 = 0.1052$, $(1-\delta_n) = 1 - 0.1052 = 0.8948$

$\sum G_i H_i = 1885.50 \times 5.5 + 1869.38 \times 10 +$
$\qquad 1866.98 \times 14.5 + 1244.14 \times 19$
$\qquad = 79,773.9$ kN · m

$$F_1 = \frac{1885.50 \times 5.5}{79,773.9} \times 245.12 \times 0.8948$$
$$= 28.50 \text{ kN}$$
$$F_2 = \frac{1869.38 \times 10}{79,773.9} \times 245.12 \times 0.8948$$
$$= 51.40 \text{ kN}$$
$$F_3 = \frac{1866.98 \times 14.5}{79,773.9} \times 245.12 \times 0.8948$$
$$= 74.43 \text{ kN}$$
$$F_4 = \frac{1244.14 \times 19}{79,773.9} \times 245.12 \times 0.8948 +$$
$$0.1052 \times 245.04 = 90.77 \text{ kN}$$

3) Internal force and deformation calculation of the structure

The internal force is equal to 1.2 times the representative gravity load plus 1.3 times the earthquake action, which only one direction is considered when calculating the internal force.

The calculation of the structural internal forces under the representative value of the gravity load:

The bending moment diagram of the frame under the representative gravity load is depicted in Fig. 7-31, the shear diagram is illustrated in Fig. 7-32 and the axial force diagram is illustrated in Fig. 7-33. The lateral displacement of the steel frame nodes under only a horizontal earthquake force is given in Table 7-9.

Lateral displacement of the steel frame nodes under horizontal earthquake force

Table 7-9

Node	Lateral displacement (mm)	Node	Lateral displacement (mm)	Node	Lateral displacement (mm)	Node	Lateral displacement (mm)
1	0	6	0	11	0	16	0
2	9.21	7	9.16	12	9.11	17	9.05
3	18.24	8	18.08	13	17.99	18	17.96
4	25.23	9	25.01	14	24.88	19	24.83
5	29.34	10	29.06	15	28.90	20	28.85

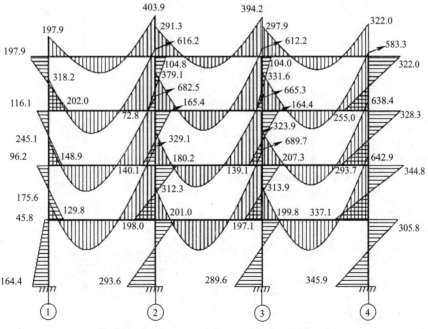

Fig. 7-31 Bending moment diagram of the frame (kN · m)

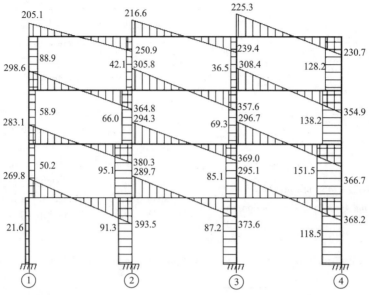

Fig. 7-32 Shear diagram of the frame (kN)

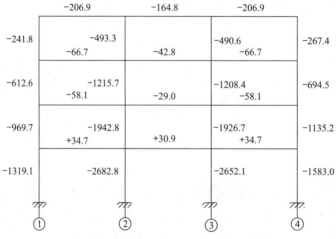

Fig. 7-33 Axial force diagram of the frame (kN)

2. Earthquake check

The height of the frame is at most 50 m; thus, the seismic intensity is 7 and the seismic grade is 4 based on the ***Code for Seismic Design of Buildings*** (GB 50011—2010) Version 2016.

1) Check the deformation under frequently occurring earthquakes

The elastic storey drift of each floor is as follows:

$$\Delta_{ue1} = \frac{8.19+8.16+8.12+8.07}{4} = 8.14 \text{ mm}$$

$$\Delta_{ue2} = \frac{15.77+15.65+15.58+15.56}{4} -$$

$$\frac{8.19+8.16+8.12+8.07}{4}$$

$$= 7.51 \text{ mm}$$

$$\Delta_{ue3} = \frac{21.97+21.8+21.7+21.65}{4} -$$

$$\frac{15.77+15.65+15.58+15.56}{4}$$

$$= 6.14 \text{ mm}$$

$$\Delta_{ue4} = \frac{25.61+25.4+25.27+25.23}{4} -$$

$$\frac{21.97+21.8+21.7+21.65}{4}$$

$$= 3.60 \text{ mm}$$

The allowable value of the drift angle is:

$[\theta_e] = \dfrac{1}{250}$

$\Delta_{ue1} = 8.14$ mm $< [\theta_e]h_1 = 22.0$ mm

$\Delta_{ue2} = 7.51$ mm $< [\theta_e]h_2 = 18.0$ mm

Therefore, this structure conforms to the deformation requirements.

2) Calculation of the strength and stability check (we use axes 1 and 2 of FL2 as an example, and the detailed section size is depicted in Fig. 7-34)

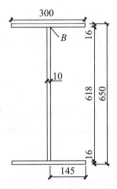

Fig. 7-34 Section of L1 (mm)

We can obtain the internal force from the structure analysis:

$M'_b = 686.3$ kN, $V'_b = 382.0$ kN, $N'_b = 66.4$ kN

The internal force for the design is the design value at the end of the beam:

$V_b = 380.0 - (281.4 + 382.0) \times \dfrac{0.25}{9}$

$\quad = 363.6$ kN

$M_b = 683.3 - \left[363.6 \times 0.25 + (382.0 - 363.6) \times \dfrac{0.25}{2}\right]$

$\quad = 593.1$ kN·m

$N_b = 66.4$ kN

Beam section properties: $A_n = 157.8$ cm^2, $W_x = 3574.1$ cm^3.

Design strength of the steel beam: $f_b = 215$ MPa, $f_{vb} = 125$ MPa, $\gamma_{RE} = 0.75$.

(1) Cross-sectional normal stress check

$\dfrac{N_b}{A_n} + \dfrac{M_b}{\gamma_x W_x} \leq \dfrac{f_b}{\gamma_{RE}}$

Plug in the data:

$\dfrac{66.4 \times 10^3}{157.8 \times 10^2} + \dfrac{593.1 \times 10^6}{1.0 \times 3574.1 \times 10^3}$

$= 162.2$ MPa $< \dfrac{f_b}{\gamma_{RE}} = \dfrac{215}{0.75} = 286.7$ MPa

Thus, it meets the requirement.

(2) Cross-sectional shear stress check

$\tau_b = \dfrac{V_b}{A_w} = \dfrac{363.6 \times 10^3}{6180} = 58.8 \leq \dfrac{f_{vb}}{\gamma_{RW}}$

$= \dfrac{125}{0.75} = 166.67$ MPa

Thus, it meets the requirement.

(3) Cross-sectional complex stress check

$\sigma_B = \dfrac{N_b}{A_n} + \dfrac{M_b}{\gamma_x M_x} \cdot \dfrac{h_B}{h} = \dfrac{66.4 \times 10^3}{157.8 \times 10^2} + \dfrac{593.1 \times 10^6}{1.0 \times 3574.1 \times 10^3} \times \dfrac{325 - 16}{325}$

$= 154.5$ MPa

$\tau_B = 47.6$ MPa

$\sqrt{\sigma_B^2 + 3\tau_B^2} = \sqrt{154.5^2 + 3 \times 47.6^2}$

$= 175.1$ MPa $< \dfrac{215}{0.75}$

$= 286.7$ MPa

(4) Stability of the beam check

The beams have enough stiffness at the ceiling; thus, we can ignore the calculation for the overall stability.

Flange:

$b/t = 145/16 = 9.1 < 11$

(According to Table 7-5)

Web:

$\dfrac{N_b}{A_f} = \dfrac{66.4 \times 10^3}{157.8 \times 10^2 \times 215} = 0.020$

$85 - \dfrac{120 N_b}{A_f} = 82.7 > 75$

Take the flakiness ratio of web as 75, then:

$\dfrac{h}{t} = \dfrac{618}{10} = 61.7 \leq 75$

(According to Table 7-5)

Thus, it meets the strength and stability requirements.

3) Strength and stability check for the columns (use the upper columns along axis 2 as examples, and the section size is indicated in Fig. 7-35)

Internal force design values:

$M_c = 292.1$ kN·m, $V_c = 89.8$ kN, $N_c =$

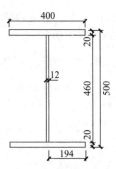

Fig. 7-35 Section of Z1 (mm)

2690.5 kN

Column section features:
$A = 215.2 \text{ cm}^2$, $W_x = 4077.9 \text{ cm}^3$

Column design strength:
$f_c = 295 \text{ MPa}$

Strength coefficient:
$\gamma_{RE} = 0.75$

Stability coefficient:
$\gamma_{RE} = 0.8$

(1) Calculation of the gravity additional bending moment M_g and the primary bending moment M_0 of each floor

$$M_{gi} = \Delta_{uei} \sum N_{ik}$$

where,

$\sum N_{ik}$——the sum of the standard pressure values for the axes on each floor.

$$M_{0gi} = hi \sum F_k$$

where,

k——the sum of the shear force for each floor.

First floor:
$M_{g1} = (1205.3 + 2227.7 + 2227.7 + 1205.3) \times$
$\quad 8.14 \times 10^{-3} = 55.9 \text{ kN} \cdot \text{m}$
$M_{01} = (90.77 + 74.43 + 51.40 + 28.50) \times$
$\quad 5.5 = 1340.1 \text{ kN} \cdot \text{m}$

Second floor:
$M_{g2} = (873.3 + 1617 + 1617 + 873.3) \times$
$\quad 7.51 \times 10^{-3} = 37.4 \text{ kN} \cdot \text{m}$
$M_{02} = (90.77 + 74.43 + 51.40) \times 4.5$
$\quad = 974.7 \text{ kN} \cdot \text{m}$

Third floor:
$M_{g3} = (542 + 1013.7 + 1013.7 + 542) \times$
$\quad 6.14 \times 10^{-3} = 19.1 \text{ kN} \cdot \text{m}$
$M_{03} = (90.77 + 74.43) \times 4.5$

$\quad = 743.4 \text{ kN} \cdot \text{m}$

Fourth floor:
$M_{g4} = (201.7 + 411.4 + 411.4 + 210.7) \times$
$\quad 3.60 \times 10^{-3} = 4.5 \text{ kN} \cdot \text{m}$
$M_{04} = 90.77 \times 4.5 = 408.5 \text{ kN} \cdot \text{m}$

On each floor $M_{gi} < 0.1 M_{0i}$; thus, we can ingore the secondary effect of gravity.

(2) Cross-sectional normal stress of the column check

$$\frac{N_c}{A_n} + \frac{M_c}{\gamma_x W_x} \leq \frac{f_c}{\gamma_{RE}}$$

Plug in the data:

$$\frac{2690 \times 10^3}{215.2 \times 10^2} + \frac{292.1 \times 10^6}{1.05 \times 4077.9 \times 10^4}$$

$= 131.8 \text{ MPa} < \dfrac{295}{0.75}$

$= 393.33 \text{ MPa}$

Thus, it meets the requirement.

(3) Calculation to check the stability of the columns

Based on the code for steel structure design, we can obtain the calculation factors shown in Fig. 7-36.

1.647	1.374	1.374	1.647
1.849	1.506	1.506	1.849
1.877	1.521	1.521	1.877
1.411	1.267	1.267	1.411

Fig. 7-36 Effective length coefficient of the columns

Overall stability:

We can determine that $\mu = 1.267$ from the Fig. 7-36.

$l_0 = \mu l = 1.267 \times 5.5 = 7.0 \text{ m}$

$\lambda = l_0/i = \dfrac{700}{\sqrt{\dfrac{101,946.9}{215.2}}}$

$= 32.2 < 120\sqrt{235/f_y}$

$= 120 \times \sqrt{235/345} = 99.0$

From the ***Design Code for Steel Structures*** (GB 50017—2017):

$\varphi_x = 0.944$

$$N_{Nx} = \frac{\pi^2 EA}{\gamma_R \gamma_x^2} = \frac{\pi^2 \times 2.06 \times 10^5 \times 215.2 \times 10^2}{1.111 \times 32.32 \times 10^3}$$
$$= 38,362.3 \text{ kN}$$

$$\beta_{mx} = 0.65 - 0.35 \times \frac{201.9}{292.1} = 0.41$$

$$\frac{2690.5 \times 10^3}{0.944 \times 215.2 \times 10^2} + \frac{0.41 \times 292.1 \times 10^6}{1.05 \times 4077.9 \times 10^3 \times \left(1.0 - 0.8 \times \frac{2690.5}{38,362.3}\right)} = 162.1 \text{ MPa}$$

$$\leqslant \frac{295}{0.8} = 368.75 \text{ MPa, meet the requirement.}$$

Local stability:

Flange:
$$b/t = (200-6)/20 = 9.7 < 13\sqrt{235/345}$$
$$= 10.7 \quad (\text{According to Table 7-5})$$

Web:
$$\frac{h}{t} = \frac{500 - 2 \times 20}{12} = 38.3 < 52\sqrt{235/f_y}$$
$$= 52 \times \sqrt{235/345} = 42.9$$
(According to Table 7-5)

Thus, the strength and stability meet the requirements.

4) Plastic bearing capacity check for the beam end and column joint (use joint 7 in Fig. 7-28 as an example)
$$\sum W_{pc}(f_{yc} - N/A_c) \geqslant \eta \sum W_{pb} f_{yb}$$
(According to Eq. 7-6a)

where,
$$W_{pc} = 2 \times \left[400 \times 20 \times \frac{1}{2} \times (500-20) + \frac{460}{2} \times 12 \times \frac{460}{4}\right]$$
$$= 4474.8 \text{ cm}^3$$

$$W_{pb} = 2 \times \left[300 \times 16 \times \frac{1}{2} \times (650-16) + \frac{+618}{2} \times 10 \times \frac{618}{4}\right]$$
$$= 3998.8 \text{ cm}^3$$

$f_{yc} = 345$ MPa, $\eta = 1.0$, $f_{yb} = 235$ MPa

Plug in the data:
$$\left[4474.8 \times 10^3 \times \left(345 - \frac{2690.5 \times 10^3}{215.2 \times 10^2}\right) + 4474.8 \times 10^3 \times \left(345 - \frac{1950.2 \times 10^3}{215.2 \times 10^2}\right)\right] \times 10^{-6}$$
$$= 2122.6 \text{ kN} \cdot \text{m}$$

Thus, it meets the requirement.

5) Yield capacity of the joint check (use joint 7 in Fig. 7-28 as an example)

Where the reduction factor is $\psi = 0.6$ (seismic Grade 4):
$$M_{pb1} = M_{pb2} = M_{pb} f = 3998.0 \times 10^3 \times 235$$

$$\frac{N_c}{\varphi x A} + \frac{\beta_{mx} M_c}{\gamma_x W_x \left(1 - 0.8 \frac{N}{N_{Ex}}\right)} \leqslant \frac{f_c}{\gamma_{RE}}$$

Plug in the data:
$$= 939.5 \text{ kN} \cdot \text{m}$$
$$h_b = 650 - 2 \times 16 = 618 \text{ mm}$$
$$h_c = 500 - 2 \times 20 = 460 \text{ mm}$$
$$t_w = 12 \text{ mm}$$
$$V_p = 618 \times 460 \times 12 = 3411.4 \text{ cm}^3$$

Plug in the data:
$$\frac{0.6 \times 2 \times 939.5 \times 10^6}{3411.4 \times 10^3} = 330.5 \text{ MPa} > \frac{4}{3} f_{yv}$$
$$= \frac{4}{3} \times 0.58 \times 345$$
$$= 266.8 \text{ MPa}$$

Thus, it does not meet the requirement.

Therefore, the node area needs to be locally thickened. We can use $t_w = 20$ mm, and then
$$V_p = 618 \times 460 \times 20 = 5685.6 \text{ cm}^3$$
$$\frac{0.6 \times 2 \times 939.5 \times 10^6}{5685.6 \times 10^3} = 198.3 \text{ MPa} < \frac{4}{3} f_{yv}$$
$$= \frac{4}{3} \times 0.58 \times 335$$
$$= 259.1 \text{ MPa}$$

Thus, it meets the requirement.

(1) Local stability

Plug the dimensions into Eq. (7-10a) to obtain:
$$t_w = 20 \text{ mm} > \frac{618 + 460}{90} = 12.0 \text{ mm}$$

Thus, it meets the requirement.

(2) Shear strength

$M_{b1} = 686.3$ kN · m, $M_{b2} = -330.4$ kN · m, $\gamma_{RE} = 0.75$

Plug them into Eq. (7-10b):
$$\frac{683.6 \times 10^6 - 330.4 \times 10^6}{5685.6 \times 10^3} = 62.6 \text{ MPa} < \frac{4}{3} f_{yv}$$
$$= \frac{4}{3} \times \frac{170}{0.75}$$
$$= 302.2 \text{ MPa}$$

Thus, it meets the requirement.

Chapter 8
Seismic Design of Non-Structural Elements

8.1 Introduction

Non-structural elements in the seismic design of construction engineering typically contain two types. The first type is fixed components attached to the supporting structure, including non-bearing walls, elements attached to the floor and roof, decorative elements, large storage racks fixed on the floor, etc. These components are typically referred to as architectural non-structural elements. The second type includes mechanical and electrical components related to construction functions, such as lighting and emergency power, elevators, communications equipment, pipeline systems, air conditioning systems, fire detection and fireproof systems and communal aerial systems.

The seismic design of non-structural elements involves several areas of expertise, and it is typically performed by designers in construction design, interior design, construction equipment, etc. For example, there are specific design standards for curtain walls and elevators as well as product standards and design requirements for mechanical and electrical equipment inside a building. Certain important equipment must be tested against their seismic performance so that they can provide seismic safety and reliability under all operating conditions. Therefore, in the seismic design of mechanical and electrical equipment for buildings, it is important to ensure that the supporting systems and the connections between these systems and the main structures are safe and reliable and meet the service life requirements. For non-structural elements, it is also essential to ensure the seismic safety of the elements themselves in addition to the safe and reliable connections between them and the main structures.

The engineering field has always focused on the seismic design of non-structural elements. In previous editions of the *Code for Seismic Design of Buildings* (GB 50011—2010) Version 2016, there were seismic design requirements for different types of construction, such as masonry buildings, multi-storey and high-rise reinforced concrete buildings, single-storey reinforced concrete columns workshops, single-storey brick column workshops and single-storey steel structure workshops. In the latest code 2016 version, the above regulations are integrated into basic requirements based on the material, type, arrangement and anchoring for architectural non-structural elements. The seismic requirements for certain non-structural elements (e.g., cladding, curtain walls, advertising boards, mechanical and electrical equipment, etc.), which focus on the integrity of these elements, are not included in the *Code for Seismic Design of Buildings* (GB 50011—2010) Version. Additionally, the production equipment and relevant facilities in industrial buildings are not included in the equipment for the building, which should be designed separately.

The objective of seismic fortification of non-structural elements should be consistent with the "two-stage and three-level" method of seismic design of the primary structural system, in which the damage of non-structural elements is allowed to be slightly greater than that of the primary structure as long as it does not pose a threat to human lives.

High requirements: The appearance may be damaged but the damage does not affect the function of serviceability and fire protection, the facilities and emergency system can function normally, the deformation of the connected structural elements can be 1.4 times the design deflection of the architectural non-structural elements or equipment support, which means that the non-structural elements have high adaptability for deformation.

Medium requirements: Facilities can be maintained or repaired, fire resistance is reduced by 25%, the emergency system can operate normally, can undergo the deformation of connected structural elements equivalent to 1.0 times the design deflection of the architectural non-structural elements or equipment support.

General requirements: The fire resistance period has obviously decreased, the equipment can operate normally after repair, the emergency system can operate normally while damaged, the deformation of the connected structural elements can be 0.6 times the design deflection of the architectural non-structural elements or equipment support.

The above requirements should be calculated based on the seismic fortification classification, the seismic damage consequences of non-structural elements and their impact on the overall structure with appropriate seismic measures, functional coefficients and category coefficients to ensure compliance with the requirements through both seismic measures and seismic calculations. When two non-structural elements are connected, the seismic design should be conducted based on the one with the higher seismic requirement. When the connection between two non-structural elements is damaged, the non-structural elements with the higher seismic requirement connected to them should not be disabled.

8.2 Key Points for Seismic Design

Seismic damage to non-structural elements is extremely common. 60% of the seismic codes of the major countries around the world require seismic impact calculations for non-structural elements, whereas only 28% specify requirements for the seismic detailing of non-structural elements. The earlier 1989 version of the seismic code of China had requirements for the seismic calculation of parapets and their long cantilever structures (e. g., canopies). The *Code for Seismic Design of Buildings* (GB 50011—2010) Version 2016 specifies that the impact of non-structural elements should be included in the seismic calculation of the structural system. It also specifies the basic calculation method for the seismic impact of non-structural elements.

8.2.1 Influence of Non-Structural Elements on Main Structure Calculation

The impact of non-structural elements can be included in the construction calculation for the seismic action of a structure as follows:

(1) Non-structural elements should consider the seismic action caused by their gravity. Seismic action calculations should take into account the weight of architectural members and mechanical and electrical equipment attached to structural members should be included.

(2) In general, the positive effect of non-structural elements is left out, such as masonry infilled walls, which contribute to the lateral stiffness of the structure. For construction structures in which the primary structure uses flexible connections, the rigidity can be excluded. For non-structural elements with high rigidity embedded in the in-plane lateral force-resistant element, their stiffness influence can be accounted for using simplified methods, such as adjusting the structural system period. The seismic bearing capacity of these non-structural elements is typically not included unless there are specific structural measures. For example, if the in-filled masonry walls in the frame structure have a reliable connection to the frame columns or beams, and the seismic structure of such infilled masonry wall complies with the lateral force-resistant masonry wall requirements, then its positive impact should be considered in the seismic calculation.

(3) Electromechanical equipment should

consider collaboration with structure by using floor response spectrum, for the mechanical and electrical equipment, which requires floor spectra calculations, a proper and simplified model should be used to include the interaction between the equipment and the structure. In the calculation of the floor spectra, a single-degree-of-freedom (SDOF) dynamic model can be used for non-structural elements; for non-structural elements with a relative displacement between the supports, it is better to use a multi-degree-of-freedom (MDOF) dynamic model to perform the calculation.

(4) For major structural elements that support non-structural elements, the seismic impact of the non-structural elements should be added as a force to the major structural elements, and the connection should comply with the anchoring requirement.

The above regulations identify the mechanical relationship betweenthe non-structural elements and the primary structure. The force that non-structural elements apply to the primary structure should be added to the calculation based on the actual conditions, and the design or verification of the element's section should be performed accordingly.

8.2.2 Seismic Calculation of Non-Structural Elements

(1) Using the equivalent lateral force method, the forces act at gravity center. The seismic force of non-structural elements should be applied to their centers of gravity, whereas the horizontal seismic force can be applied in any horizontal direction.

(2) The seismic impact caused by the gravity of non-structural elements can be calculated using the equivalent lateral force method; for non-structural elements supported on a different storey or at both sides of a seismic joint, the impact of the relative displacement between the support points during an earthquake should be considered in addition to the gravity of the non-structural elements. This impact can be calculated by multiplying the lateral stiffness in the displacement direction with a horizontal displacement of the support point. The stiffness of the non-structural elements in the displacement direction should use simplified mechanical models, such as rigid, hinged, elastic or slip connections, based on the actual end connections characteristic. The relative displacement of the adjacent storeys should use the threshold regulated in the ***Code for Seismic Design of Buildings*** (GB 50011—2010) Version 2016; the relative displacement at both sides of the seismic joints can be determined based on the relevant requirements. For example, if a pipeline or support is connected with devices on adjacent storeys, in addition to the inertia force of the pipeline or support, the force applied by the relative displacement of the two different storeys should also be considered and calculated for the worst conditions. Furthermore, if the pipeline or support spans both sides of a seismic joint, then the force applied by the displacement difference of both sides of the seismic joint should be considered.

(3) If the natural period of equipment for the building (including support) is longer than 0.1s and the gravity of the equipment exceeds 1% of the gravity of the storey (or the gravity of the equipment exceeds 10% of the gravity of the storey), then it is better to use either the seismic design of the whole structure model or the floor response spectrum to calculate the seismic impact of the equipment. If the equipment is connected to the floor system with an inelastic connection, the equipment and the floor system can be combined as one mass point in the analysis of the overall structure, and the seismic impact on the equipment can be allocated based on its weight ratio in the mass point. For example, the impact on a huge high-level water tank, large tower support that extends above the roof, a large fan and a cooling tower, and its support on the roof can be calculated using the floor response spectrum method or the time-historey analysis method.

8.2.3 Requirements for the Equivalent Lateral Force Method

When using the equivalent lateral force method, the standard value of the horizontal seismic effect can be calculated as follows:

$$F = \gamma \eta \xi_1 \xi_2 \alpha_{max} G \qquad (8\text{-}1)$$

where,

F——the characteristic value of seismic action;

γ——the function coefficient of non-structural elements, depending on the seismic category and operating requirements, generally, is 1.4, 1.0 and 0.6, as indicated in Tables 8-1 and 8-2;

η——the category coefficient of non-structural elements, depending on materials, generally, is 0.6 ~ 1.2, as indicated in Tables 8-1 and 8-2;

ξ_1——the status coefficient: 2.0 for precast members, cantilever members, any equipment braced to a structural member lower than its center of mass and flexible systems, and 1.0 for other cases;

ξ_2——the position coefficient is 2.0 at top, 1.0 at bottom, linear interpolation between top and bottom, and for structures that need to use the time-historey analysis method, the value should be adjusted based on the results;

α_{max}——the maximum value of the seismic influence coefficient, which may be determined based on the provisions for minor earthquakes;

G——the gravity of non-structural elements, including the gravity of people, media in vessels and pipelines and items in lockers during operation.

8.2.4 Calculation Requirement for the Floor Response Spectrum Method

The non-structural elements for which the floor response spectrum calculation is needed are primarily the mechanical and electrical equipment for the building. The floor response spectrum calculation can reflect the counter-effect of non-structural elements on the structure. This

Function coefficients and category coefficients for architectural non-structural elements   Table 8-1

Element	Function coefficient γ		Category coefficient η
	Category B building	Category C building	
Non-bearing exterior wall:			
Enclosure wall	1.4	1.0	0.9
Glass curtain wall, etc.	1.4	1.4	0.9
Connector:			
Wall connector	1.4	1.0	1.0
Finish connector	1.0	0.6	1.0
Fireproof ceiling connector	1.0	1.0	0.9
Non-fireproof ceiling connector	1.0	0.6	0.6
Secondary elements:			
Sign and advertising board, etc.	1.0	1.0	1.2
Support of locker higher than 2.4 m:			
Storage rack (container), file cabinet	1.0	0.6	0.6
Cultural relic cabinet	1.4	1.0	1.0

Function coefficients and category coefficients for mechanical and electrical components   Table 8-2

Element or system	Function coefficient γ		Category coefficient η
	Category B building	Category C building	
Emergency power's controlling system, generators, refrigerators, etc.	1.4	1.4	1.0
Elevators' bracing, guide rails, brackets, car guide components, etc.	1.0	1.0	1.0
Suspended or swing lamps	1.0	0.6	0.9
Other lamps	1.0	0.6	0.6
Support for cabinet equipment	1.0	0.6	0.6
Support for water tank, cooling tower	1.0	1.0	1.2
Support for boiler, pressure vessel	1.0	1.0	1.0
Support for public antenna	1.0	1.0	1.2

counter effect can change not only the seismic reaction of the structure but also those of the non-structural elements that are fixed on the structure.

There are two basic methods used to calculate the floor spectrum: time-history analysis and random vibration. If the material of a non-structural element is the same as that of the primary structure, then the floor spectrum can be directly calculated by entering multiple groups of seismic waves into a time-history analysis software program. If a non-structural element has a high mass, apparent different damping characteristics or a supporting point at a different storey, then the calculation should be checked using the second-generation floor spectrum. In this case, the interaction between the non-structural element and the primary structure may be considered, including the "effect of vibration absorption" so that the calculation result is more reliable. The calculation of the floor spectrum through the time-historey analysis and random vibration methods requires specific software. A simplified method is used in the *Code for Seismic Design of Buildings* (GB 50011—2010) Version 2016, in which the standard value of the horizontal seismic effect can be calculated as follows:

$$F = \gamma \eta \beta_s G \qquad (8-2)$$

where,

β_s——the floor response spectrum value of non-structural elements depending on intensity, site condition, period ratio of non-structural elements to structural elements, mass ratio, damping, the number and position of supports and connection condition, which reflect the dynamic characteristic of the main structure, filtering action, amplified action and attenuation of ground motion. Generally, non-structural elements are simplified as a SDOF system supported by the primary structure. If there is a relative displacement between the supports, a MDOF system and specific program should be used for the calculation.

The floor response spectrum is different from the ground response spectrum used in the structure design. It reflects the dynamic characteristics of the primary structure that supports non-structural elements, storey position of the non-structural elements, filtering or amplification effects of the primary structure on the earthquake ground motion and attenuation effects of the damping characteristics of the primary structure and non-structural elements on the earthquake ground motion. The value of the floor response spectrum must be calculated using a specific program. The obtained result can be used in Eq. (8-2) for the seismic calculation of non-structural elements.

8.2.5 Combination of Seismic Action Effect

The combination of the seismic effects of non-structural elements (including effects caused by gravity and the relative displacement of the supports) and other load effects can be calculated based on the same requirements for the combined seismic effect and other load effects of the primary structure. The combination of the seismic effect and the wind load effect should be calculated for the curtain walls, whereas the effects caused by the operating temperature and working load should be calculated for the vessels.

When performing a seismic checking calculation of non-structural elements, friction should not be considered as a resistance to the seismic effect; the seismic coefficient of the bearing capacity for the connectors should be set to 1.0.

The deformation capacity of decorative non-structural elements may vary significantly. Masonry non-structural elements are forbidden in high-requirement situations due to their poor deformability. The codes of foreign countries only regulate detailing requirements with no seismic calculation requirements on non-structural masonry elements; metal curtain walls and certain premium decorative materials have better deformability, and overseas factories typically provide these materials based on the deformation requirement of the primary structure instead of non-structural elements. For glass curtain walls, it is regulated in the ***Curtain Wall for Building*** (GB/T 21086—2007) that the in-plane deformation is divided into five classes, with the highest being 1/100 and the lowest being 1/400.

8.3 Basic Seismic Measures of Architectural Non-Structural Elements

Referring to the ***Code for Seismic Design of Buildings*** (GB 50011—2010) Version 2016, the primary measures for non-structural elements and regulations for their design are as follows:

(1) Requirements about connectors and connected positions

Certain locations in buildings where embedded parts and anchoring parts are used to connect non-structural elements, such as curtain walls, enclosures, partition walls, parapets, awnings, signs, billboards, ceilings and large-size storage racks, should be strengthened to resist the seismic effects transferred by the non-structural elements to the primary structure. For example, locations with these embedded parts on reinforced concrete elements are generally intensified by stirrups; the anchor bars of these embedded parts should penetrate sufficiently into the elements or should be reliably connected to the main bars. The concrete around the embedded parts should have suitable vibration abilities to prevent it from being less dense under the embedded parts, i.e., having holes.

(2) Requirements about materials, types and layouts of non-bearing walls

The material, type and arrangement of non-bearing walls should be determined through an overall analysis, including factors such as intensity, building height, building shape, interstorey drift and lateral force resisting capacity of the walls.

(a) Wall materials should be selected based on the following requirements: non-bearing walls should use lightweight materials. If selecting a masonry wall, measures should be taken to reduce the adverse effects on the primary structure, and steel tie-bars, horizontal tie-bands, ring beams and tie-columns should be used to ensure reliable connections to the primary structure.

(b) The layout of rigid non-bearing walls should avoid a sudden change in the rigidity and strength distribution on the structure. The rigid enclosures of a single-storey factory building with reinforced concrete columns should be set uniformly in the longitudinal direction instead of being set outside on one side and embedded on the other or in an open manner. Furthermore, it is improper to use masonry walls on one side and lightweight boards on the other. If the enclosures are not uniformly arranged, the adverse seismic impacts on the primary structure caused by the difference in mass and rigidity should be considered.

(c) Reliable ties should be used between the walls and the primary structure so that they can adapt to the inter-storey displacement of the primary structure in different directions. They should possess deformability to adapt to the inter-storey displacement at intensities 8 and 9. Additionally, when connected to the overhanging elements, they should adapt to the vertical deformation caused by the joint rotation.

(d) The connectors of the external wall panels should possess sufficient ductility and proper rotational ability to meet the inter-storey displacement of the primary structure under certain seismic intensity. For example, if the maximum allowed elastic inter-storey displacement of the primary structure occurs, the connectors of the external wall panels should possess an elastic deformability; if the maximum allowed elastic-plastic deformation of the primary structure occurs, then the connectors of the external wall panels should not be damaged to the extent of the failure, and they should ensure that large portions of the panels do not fall off.

(e) Masonry parapets should be anchored to the primary structure at the entrances and exits of the building; the height of the parapets where there are no entrances and exits and without anchorage should not exceed 0.5 m at intensities 6 to 8 and should be anchored at intensity 9. Parapets at seismic joints should have sufficient width, whereas the free ends at both sides of the joints should be strengthened.

(3) Requirements about non-bearing masonry walls

Measures should be taken to reduce the adverse impacts of non-bearing masonry walls on the primary structure, whereas steel tie-bars, horizontal tie-bands, ring beams and tie-columns should be set to ensure reliable connections with the primary structure.

(a) A non-bearing wall in a multi-storey masonry structure should comply with the following requirements: ① The post-laying non-bearing partition walls should have $2\phi6$ steel tie-bars every 500—600 mm in the vertical to connect to the load-bearing wall or the column and be extended into each wall with a length of at least 500 mm; for intensities 8 and 9, when the length of the post-laying partition walls exceeds 5 m, the top of these walls should be tied together with the slabs or beams, and the separate wall end and large door edge should have reinforced concrete tie-columns. ② The flue, air flue and rubbish chute should not weaken the walls; if the walls are weakened, strengthening measures should be taken. It is improper to use a wall-attached chimney without vertical bars or a chimney outreaching the roof. ③It is improper to use a reinforced concrete prefabricated cornice without anchorage.

(b) Masonry in-filled walls in reinforced concrete structures should be separated from the columns or use a flexible connection with the columns; furthermore, they should comply with the following requirements: ①The arrangement of the filling wall in the plane and vertical layouts should be symmetric and even, i.e., weak storeys or short columns should be avoided. ②The mortar strength of the masonry should be at least M5; the strength level of the solid brick should be at least MU 2.5, and the strength level of the hollow block should be at least MU 3.5. The wall top should be closely joined with the frame beams. ③The in-filled walls should be set with 2ϕ6 tie bars every 500—600 mm along the full height of frame column, the length extended into the wall, may be overall length of wall for intensities 6 and 7 and should be the overall length for intensities 8 and 9. ④If the length of a wall exceeds 5 m, then the top should be tied to the corresponding beam; if the length of the wall exceeds 8 m or twice the storey height, a reinforced concrete tie column should be built. Furthermore, if the height of the wall exceeds 4 m, then a horizontal tie-band should be installed at the half height of the wall and connected to the columns at both ends of the wall. ⑤In-filled walls in staircases and passages should be strengthened using steel wire mesh-reinforced mortar.

(c) Masonry partition walls and enclosures of single-storey factory buildings with reinforced concrete columns should comply with the following requirements: ①The masonry partition walls should be separated from the columns or use a flexible connection to the columns; effective measures should be taken to ensure the stability of the walls, and cast-in-situ reinforced concrete coping should be set on the top of the partition walls. ②The masonry enclosure should be laid to close the outside and tie reliably to the columns; the fascia wall above the lower roof of factory buildings with different heights and the suspended walls at the intersections of the longitudinal and transversal units should use lightweight wall panels, and for intensities 6 and 7, the masonry walls should not be directly laid on the lower roof. ③The masonry enclosure should be set with a cast-in-situ reinforced concrete ring beam at the following locations. The ring beam should be installed at the levels of the top chord of the trapezoid trusses and the top of the columns; however, if the height of the truss end posts is not greater than 900 mm, the ring beam may be synthetically installed. The ring beam should be placed at every 4 m intervals on the top of the window based on the principle that the vertical spacing of the ring beam in the upper portion should be denser than that in the lower portion. Furthermore, for the fascia walls above the lower roof of factory buildings with different heights and the suspended walls at the intersections of the longitudinal and transversal units, the vertical distance between each beam should not exceed 3 m. A ring beam should be installed on the gable walls along the roof level, and the beams should be connected to the ring beam at the top chord level of the truss ends. ④The details of the ring beam should comply with the following requirements. The ring beam should be enclosed with a section width equivalent to the wall thickness, and the beam depth should be at least 180 mm. Additionally, the longitudinal reinforcement should be at least 4ϕ12 for intensities 6 to 8 or 4ϕ14 for intensity 9. For the column top ring beam at the corners of factory units, the longitudinal reinforcement at the end bay should be at least 4ϕ14 for intensities 6—8 or 4ϕ16 for intensity 9. The diameter of the stirrups within 1 m on two sides of the corners should be at least ϕ8, and the spacing should not exceed 100 mm. At least 3 horizontal diagonal bars should be added to the corner of the ring beam with the same diameter as the longitudinal bars at the corners of the ring beam; the ring beam should be firmly connected with the column or the roof truss, and the ring beam of the gable wall should be tied to the roof slabs. The anchoring bar that connects the top ring beam with the columns or roof trusses should be at least 4ϕ12, with an anchorage length of at least 35 times the diameter of the steel bars, and the connection between the ring beam and the columns or roof trusses at the seismic joints should be intensified. ⑤ For site Classes Ⅲ and Ⅳ of intensity 8 or 9, the precast foundation beams of the brick enclosure should use cast-in-situ joints; when a stripped footing of the enclosure wall is established, continuous cast-in-situ reinforced concrete ring beams should be installed at the top of the column footing with bars of at least 4ϕ12. ⑥The wall beam should be cast-in-situ. If a precast wall beam is used, then the bottom of the beam should be firmly tied to the top of the brick walls and anchored to the columns, and the adjacent wall beams at the building corners of the factory building should be firmly connected.

(d) The masonry enclosure of a single-storey steel structure factory building should use attached construction and be tied to the columns, and certain measures should be taken so that the wall does not hinder the horizontal displacement of the columns along the longitudinal direction of the factory building; for intensities 8 and 9, the masonry enclosure should not use attached construction.

(e) Masonry parapets should not be higher than 1 m and should be prevented from falling during an earthquake using certain measures.

(4) Requirements about ceiling components and floor connectors

The connected parts of various ceilings with floors should be able to undertake the weight of the ceiling, suspended heavy objects and the relevant mechanic and electrical equipment, as well as the additional seismic action. The bearing capacity of the anchoring should be greater than the bearing capacity of the connected parts.

(5) Requirements about layouts of cantilever canopy or canopy supported by pillars at one end.

They should be reliably connected to the primary structure.

(6) Seismic design of glass curtain walls, precast walls, accessory cantilever elements and large storage rack should accord with relevant provisions.

The above items are the minimum requirements for the seismic measures of non-structural elements. Designers can use stronger and more efficient measures based on the owner's requirement or the actual conditions of the primary structure and non-structural elements to further improve the seismic capacity of non-structural elements.

8.4 Basic Seismic Measures for Mechanical and Electrical Equipment Support

The primary causes of damage to mechanical and electrical equipment and facilities as well as connecting elements and components of a structural system are as follows: (1) elevator counterweight goes off-track; (2) relative displacement of supports causes damage to pipeline connectors; (3) insecure connections between the post-cast foundation for the equipment and the primary structure or insufficient strength of the fastening bolts causes the equipment to shift or fall from the support; (4) insufficient strength of suspended elements causes the electrical lighting to fall off; and (5) unnecessary vibration isolation devices increase the frequency, which may generate resonance due to negligence of the seismic effect in the design phase, weakening the seismic capacity of the overall equipment support system.

The seismic measures for the connecting elements and components between the elevator, lighting and emergency power systems, smoke and fire detection and firefighting systems, heating and air condition systems, communication systems and the communal aerial and primary structures should be determined after a comprehensive analysis based on specific standards and factors, such as seismic intensity, building function, building height, structural type and deformation characteristic as well as the location and operating requirement of the equipment. In the building seismic design process, the basic seismic requirements for the support of the aforementioned mechanical and electrical equipment are as follows:

(1) There is no need to consider seismic requirements for supports of the following me-

chanical and electrical equipment: ①equipment lighter than 1.8 kN; ②gas pipelines with an inner diameter less than 25 mm and electrical conduit with an inner diameter less than 60 mm; ③rectangular air pipes with a cross-sectional area less than 0.38 m^2 and circular pipes with a diameter less than 0.7 m; and ④conduits suspended by a hanger with a calculated length less than 300 mm.

(2) Mechanical and electrical equipment should not be installed in locations where they may cause operation failure or other secondary damage. For equipment with isolation devices, the effects of strong vibrations on the connected parts should be considered, and resonance of the equipment with the building structure should be avoided.

The braced frames for mechanical and electrical equipment should have sufficient stiffness and strength, and they must have reliable connections and anchoring with the primary structure to ensure that the equipment can restore failure-free operation in the occurrence of an earthquake of design-based intensity.

(3) The damage to pipelines during an earthquake occurs primarily to connectors caused by the relative displacement between supports or between the equipment and corresponding support. An improper layout of the openings of pipelines, cables, air-pipes and equipment will decrease the seismic capacity of the primary load-bearing structural members. Therefore, strengthening measures should be taken around the perimeter of the openings.

For the proper design of supports and connections, measures that increase the deformability of the pipeline connectors (e.g., flexible connector) as well as those that allow a certain relative deflection between the pipelines, equipment and structure system (e.g., suspended support and sliding support) should be taken.

(4) The support and connectors for mechanical and electrical equipment should be able to transfer all of the seismic action of the equipment to the building structure. The locations for fixing embedded parts and anchoring parts for mechanical and electrical equipment should be strengthened to resist the seismic action transferred from the equipment to the primary structure.

(5) Water tanks located at the higher position of the building should be reliably connected to the structural members. The seismic effects on the structure caused by the weight of a water tank and its water should be considered.

(6) The support of the mechanical and electrical equipment (such as fire detection and firefighting systems), which must operate continuously during the earthquake of design-based intensity should be guaranteed to run normally. Heavy equipment should be installed at locations where the seismic response is smaller, and corresponding strengthening measures should be taken for the structural members in the relevant locations.

The above six items are the basic seismic requirements for equipment supports. If the equipment supports are made of different structural materials and components (such as steel components, concrete components and masonry components), then they should also meet the seismic structure measures and seismic checking calculation requirements of those materials and components. The design for such supports can be found in the relevant chapters.

8.5 Simplified Seismic Analysis Method

For seismic calculations of mechanical and electrical equipment, it is better to use the floor response spectrum method if the period of the mechanical and electrical equipment (including support) is longer than 0.1 s, the gravity of the equipment exceeds 1% of the gravity of the storey, or the gravity of the equipment exceeds 10% of the gravity of the storey. The floor response spectrum method needs specific software, yet few programs can be used for engineering applications. Therefore, a simplified method is introduced in this section, which can be used in the general structure analysis program or the structure design software to perform the necessary calculations.

8.5.1 Time-Historey Analysis Method for Calculating the Seismic Response of Equipment

For a multi-storey building with mechanical and electrical equipment installed on a certain floor, the equipment and the floor will interact with each other during an earthquake, and their seismic response can be calculated using the model outlined in Fig. 8-1. We can assume a four-storey frame structure with equipment installed on the first and third floors, as indicated in Fig. 8-1(a). The model in Fig. 8-1(b) can be used for determining the seismic response. This is a structure with 6 degrees of freedom. The equation of motion for the system can be established based on the mass and rigidity distribution for the seismic response analysis to obtain the seismic force on the equipment as well as the displacement response and acceleration response of the equipment. There are developed structure analysis programs and structure design programs for such calculations, in which the ground motion acceleration input during an earthquake can be selected based on the requirements for general construction structures in the *Code for Seismic Design of Buildings* (GB 50011—2010) Version 2016. There is no need to perform massive floor response spectrum calculations, which makes this method feasible for designers. If there are several individual pieces of equipment on a floor and the location of these equipment changes significantly (e.g., process equipment in a multi-storey industrial factory), the number of calculations may be large. Therefore, a more practical seismic calculation method will be introduced in the next section.

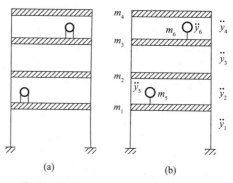

Fig. 8-1 Structural sketch and calculation model
(a) Structural sketch; (b) Calculation model

8.5.2 Practical Calculation Method for the Seismic Response of Equipment on the Floor

For an easy application of the seismic calculation method for equipment on the floor, the response acceleration of the floor is used as an equivalent acceleration for the equipment on that floor, which indicates that the system in Fig. 8-2(a) is simplified to the calculation model in Fig. 8-2(b), where the masses of all equipment are added to their corresponding floor for the calculations and the result $\ddot{y}'_1$ is the equivalent acceleration input for equipment m_5 on the first floor; whereas the result $\ddot{y}'_3$ is the equivalent acceleration input for equipment m_6 on the third floor, as indicated in Fig. 8-2(c). Then, based on the response spectrum curve and method outlined in the *Code for Seismic Design of Buildings* (GB 50011—2010) Version 2016, the seismic force on the equipment can be determined based on the equivalent acceleration input. The basis for this simplified method is as follows: (1) the weight of the equipment is extremely small compared to that of the entire floor, i.e., 1/500—1/10; thus, the effect applied to the floor by the equipment is small; (2) the frequencies of the equipment and support system are generally considerably higher than that of the building; thus, it is less possible to generate resonance for the building and the floor; (3) when using the response spectrum method to determine the seismic effect, the maximum absolute acceleration is the major controlling parameter in the case of buildings on a certain site. The response spectrum in the *Code for Seismic Design of Buildings* (GB 50011—2010) Version 2016 can be directly applied when using the maximum absolute acceleration of the floor to calculate the seismic effect of equipment on the floor, which can coordinate with the seismic calculation method of the building structure.

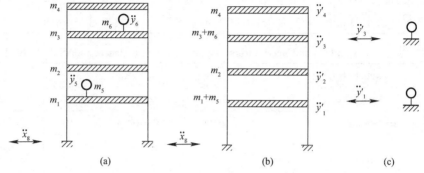

Fig. 8-2 Simplified calculation model
(a) Calculation model; (b) Simplified model; (c) Equivalent input

8.5.3 Verification of the Practical Seismic Response Calculation Method

(1) Verification based on the time-history analysis method for the seismic response

Using a four-storey frame structure as an example, process equipment is installed on the first and third floors. The simplified calculation model for the time-history analysis method to determine seismic response is depicted in Fig. 8-2(a), and the mass and rigidity are listed in Table 8-3. Based on the data in Table 8-3, the basic period of the system is 0.3875 s. The corresponding input is the G3 artificial wave, which has the same characteristics as site Class III for the seismic response calculations, i.e., using 0.2 g as the maximum acceleration peak. The peak acceleration response calculated using the six-degree model in Fig. 8-2(a) is provided in Table 8-3. The floor response accelerations $\ddot{y}'_1(t)$ and $\ddot{y}'_3(t)$ calculated using the four-degree model are used as the input for the equipment on the first and third floors, respectively, and the calculation results are provided in Table 8-4. The peak acceleration response calculated using the SDOF model and that calculated using the six-degree model in Fig. 8-2(a) are provided in Table 8-5. Based on Table 8-5, the difference between the results calculated using the simplified method and the SDOF model and the results of the six-degree model are extremely small.

(2) Verification based on a shaking-table test

In order to verify the reliability of the above calculation method, a shaking-table test is conducted for the building model with the equipment placed on the floor. A single-storey steel frame structure with a frequency of 4 Hz is used in the test. A horizontal petrochemical machine with liquid is placed on top of the steel structure. The absolute acceleration response time history of the steel frame top and the equipment is measured in the test. The calculation of the response of the equipment is conducted by taking the calculated acceleration response of the steel frame top as the input for the equipment, in which the mass and rigidity of the equipment are obtained by taking the building structure as a two-degree model. Table 8-6 provides the test and calculation results of the peak acceleration response. The calculated value and test value agree well, which means that the practical method is rational and reliable.

Parameters and acceleration response for a six-degree model   Table 8-3

Particle No.		1	2	3	4	5	6
Weight(t)		700	600	600	400	2.0	1.5
Stiffness(kN/m)		1.2×10^6	1.2×10^6	1.2×10^6	1.2×10^6	2.0×10^6	2.0×10^6
Acceleration (m/s²)	A_{min}	−2.6025	−3.7982	−4.1837	−4.1516	−2.9009	−4.2727
	A_{max}	2.7904	3.3972	4.5230	5.1825	3.1811	4.6110

Parameters and acceleration response for a four-degree model   Table 8-4

Particle No.		1	2	3	4
Weight(t)		702	600	601.5	400
Stiffness(kN/m)		1.2×10^6	1.2×10^6	1.2×10^6	1.2×10^6
Acceleration(m/s²)	A_{min}	−2.607	−3.800	−4.182	−4.148
	A_{max}	2.792	3.398	4.522	5.179

Parameters and acceleration response for a single-degree model   Table 8-5

Peak acceleration(m/s²)	Equipment reaction on first floor			Equipment reaction on third floor		
	SDOF model	6-DOF model	Error	SDOF model	6-DOF model	Error
A_{min}	−2.8126	−2.9009	3%	−4.2534	−4.2727	0.5%
A_{max}	3.0506	3.1811	4%	5.5874	4.6110	0.5%

Comparison of the shake-table test results and the calculation results   Table 8-6

Parameter	Steel frame acceleration response (g)		Floor equipment acceleration response (g)	
	A_{min}	A_{max}	A_{min}	A_{max}
Test results	−0.240	0.250	−0.511	0.438
Calculation results	−0.250	0.262	−0.540	0.495
Error(%)	4	4.8	5.7	13

Chapter 9
Introduction to Seismic Isolation and Energy Dissipation for Building Structures

9.1 Introduction

9.1.1 Seismic Isolation

The conventional seismic design takes advantage of the strength of materials and the plastic deformation capacity of structural elements to resist seismic actions and protect buildings from irreparable damage or collapse. Isolation is a special technology that can prevent superstructures from enduring seismic action, therefore, these buildings experience extremely small vibrations during earthquakes. This type of vibration does not cause damage to structures and facilities or affect the normal operation of equipment and instruments.

Engineering isolation is a branch of earthquake engineering; thus far, various types of isolation methods have been developed to meet different engineering requirements. Among the different types of isolation for building structures, the base isolation technology is fully developed and widely used. This chapter primarily introduces the base isolation technology and the design principles.

The international study of isolation technology began in the 1960s, and several buildings used this technology. After 1970, multi-layer rubber bearings were developed and applied in a few nuclear power plants in South Africa and France. In 1982, multi-layer rubber bearings with a lead core were first used in a building in New Zealand. The isolation design had been written in code for the design of buildings in several countries, providing approval and a basis for the engineering application of isolation technology.

The domestic study of isolation technology began in the 1970s, and in the 1980s, isolation technology was applied to actual projects. The Metallurgical Construction Research Institute had studied sand cushion isolation for several years and constructed a four-storey isolated building in Beijing. Since the 1990s, the study of base isolation has flourished in China. To date, more than 1000 isolated buildings have been constructed all over the country, and a few have installed vibration test instruments to obtain data during an earthquake. Although the application in China is late, base isolation has a broad potential and will be one of the development directions for the seismic design of engineering structures in the future.

9.1.2 Seismic Energy Dissipation

When an earthquake occurs, structures vibrate as a response to the ground motion. The energy absorbed by structures from an earthquake must be transformed or dissipated to end the vibration. The moderate damage of structures and bearing members (e.g., columns, beams, joints, etc.) is one form of energy dissipation, and serious damage or collapse indicates the end of energy transformation or dissipation. However, this is a passive seismic method without the capacity of self-adjustment and self-control.

A reasonable and effective seismic method is implementing a control mechanism (system) for structures. This mechanism shares the seismic action with the structure itself to suppress the structural response. In 1972, the concept of structural control was introduced by Yao J. Y. P., an American scholar. Based on the need for external power input, the vibration control of structures can be subdivided into passive control, active control, semi-active control, intelligent control and hybrid control approaches.

Energy dissipation, which belongs to engineering vibration control technology, is a type of passive control technology. In the 1970s, a few researchers, such as James M. Kelly at UC Berkeley, began to study passive energy dissipation technology. The so-called energy dissipation technology is to design certain structural members (e.g., braces and shear walls) to dissipate energy or install dampers on structures at certain locations (nodes or joints). Under minor earthquakes and wind loads, these energy-dissipating members or dampers remain elastic, and the structure has enough lateral stiffness to meet the requirements of operational use. Under moderate earthquakes, major earthquakes and strong earthquakes, the energy-dissipating members or dampers enter the inelastic state earlier than other structural members, which produces considerable damping, thus dissipating most of the energy input and consequently decreasing the dynamic response of the structure. Hence, the primary structure will exhibit no clear elastoplastic deformation and be safe throughout a strong earthquake.

An energy-dissipating structural system is a reliable vibration control system. Furthermore, based on the energy dissipation by elastoplastic deformation, it mitigates the dynamic response of a structure, which is different from the traditional seismic method. There is no need to increase a structure's strength and stiffness to resist an earthquake. Consequently, this system lessens the number of shear walls and reinforcement and diminishes the member sections; therefore, it reduces the cost of a structure. The energy dissipation of energy-dissipating members and dampers is based on the deformation of the structure; thus, for a higher and more flexible structure, the energy dissipation effect will be better.

Because of the strengths mentioned above, such as reliability, economic feasibility and wide serviceability, an energy-dissipating structural system is widely applied in the seismic and wind-resistant design for various engineering structures. This technology can be applied to not only the construction of new buildings but also the reinforcement of existing buildings and is suitable for use in both ordinary buildings and lifeline projects.

Overall, this novel technology can significantly reduce the seismic action to structures, which has been theoretically supported by existing studies and validated by the structural response.

9.2 Seismic Isolation for Building Structures

As stated above, isolation technology reduces the response of buildings exited by earthquakes and other excitation sources, such as strong winds and mechanical vibrations. Particularly, seismic isolation separates a building from its foundation, allowing it to absorb the energy of the ground motion using isolation bearings and energy dissipation devices. Generally, there are two primary types of base seismic isolation bearings, i.e., sliding friction bearings and elastomeric bearings, as indicated in Fig. 9-1.

Although the sliding form is simple, as indicated in Fig. 9-1, it is less common than elastomeric bearings due to the large displacement during powerful earthquakes. To avoid a large vertical vibration in buildings, the isolation device material requires specialized hardware that can provide large lateral displacement with vertical load capacity and recentering capabilities. Currently, multi-layer rubber bearings are widely used around the world.

In practical structures, multi-layer rubber bearings are often installed along with various types of dampers, which not only increase the natural vibration period of the structural system (usually more than 2 s) but also consequently limit energy transfer to structure and reduces both accelerations and storey drifts, thus improving the seismic behavior. The principle is described in Fig. 9-2.

Experimental results and practical phenomena have demonstrated that the seismic load and

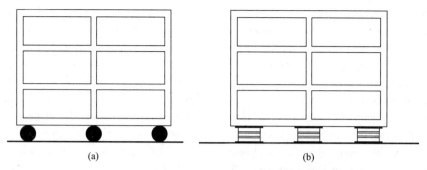

Fig. 9-1　Two types of base seismic isolation bearings
(a) Sliding friction bearings; (b) Elastomeric bearings

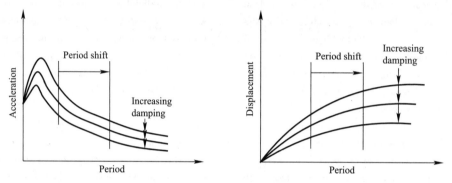

Fig. 9-2　Working principle of base seismic isolation bearings

storey drift of isolated buildings are considerably smaller than that of ordinary buildings, which is beneficial to the design of non-structural elements, building decorations and equipment tunnels.

9.2.1　Types of Multi-Layer Rubber Bearings

Multi-layer rubber bearings consist of alternating steel plates and rubber layers, and the steel plates increase the vertical stiffness and have no effect on the lateral stiffness. Three types of rubber bearings have been created that are advanced and widely applied in engineering structures.

(1) Laminated (Natural) rubber bearings. This type of bearing consists of alternating steel plates and rubber layers, which have great vertical supporting capacity.

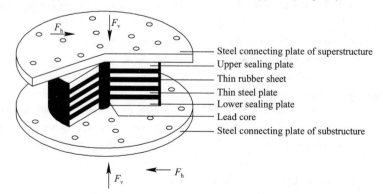

Fig. 9-3　Laminated rubber bearings with lead cores

(2) Lead rubber bearings. This type of bearing is a combination of standard laminated rubber bearings and lead cores embedded in the center of rubber layers, as indicated in Fig. 9-3. The lead cores can absorb energy and consequently reduce the deformation of buildings.

(3) Laminated rubber bearings with high damping. In this type of bearing, the rubber has high damping capabilities with great vertical stiffness and small lateral stiffness, which can remarkably absorb the energy released from earthquakes and reduce the global deformation of structures.

9.2.2 Summary

Briefly, seismic isolationcan be described by the following features:

(1) Lateral flexibility relative to the structural system.

(2) Vertically stiffness withthe ability to carry gravity load under large lateral deformations.

(3) Sufficient initial stiffness to prevent relative motion under wind and other non-seismic loads.

(4) Energy dissipation may be inherent in bearings, supplemental energy dissipation devices can be provided if not sufficient.

9.2.3 Application

The application of seismic isolation technologybegan in the 1970s. An isolation testing building in Northeastern University, Sendai City, Miyagi Prefecture, Japan, used laminated rubber bearings and hydraulic dampers and performed extremely well in an earthquake. In California, USA, Oakland City Hall laminated rubber bearings with lead cores were used for reinforcement after an earthquake. In China, the first isolated building was constructed in Shantou, which used standard laminated rubber bearings.

Some examples illustrating the importance of seismic isolation technology are shown in Fig. 9-4.

(a)

Fig. 9-4 Importance of seismic isolation technology (one)
(a) Isolated buildings

Fig. 9-4 Importance of seismic isolation technology (two)
(b) Fix-based buildings

9.3 Seismic Energy Dissipation for Building Structures

9.3.1 Characteristics of Energy Dissipation Technology

1. Principle

The principle of seismic energy dissipation technology can be described in terms of energy. The energy functions of structures during an earthquake are as follows:

For traditional seismic structures:
$$E_{in} = E_R + E_D + E_S \qquad (9\text{-}1)$$

For energy-dissipating structures:
$$E_{in} = E_R + E_D + E_S + E_A \qquad (9\text{-}2)$$

where,

E_{in}——the energy input to structures from an earthquake;

E_R——the energy of the structural response, which is the kinetic energy and potential energy of the structural vibration;

E_D——the energy dissipated by structural damping (typically at most 5%);

E_S——the energy dissipated by inelastic deformation or damage to the primary structure and load-carrying members;

E_A——the energy dissipated by the energy-dissipating members and devices.

For traditional seismic structures, the energy dissipated by structural damping E_D is neglected; thus, the structural response is eventually terminated ($E_R \rightarrow 0$) by the energy dissipation caused by damage or collapse of the primary structure and load-carrying members ($E_S \rightarrow E_{in}$).

For energy-dissipating structures, the energy dissipated by structural damping E_D is also neglected. When an earthquake occurs, the energy-dissipating members or devices dissipate most of the energy ($E_A \rightarrow E_{in}$); thus, the structural response can be rapidly reduced ($E_R \rightarrow 0$), and the primary structure and load-carrying members are protected from damage ($E_S \rightarrow 0$).

The principle can be understood using another approach, which is described in Fig. 9-5.

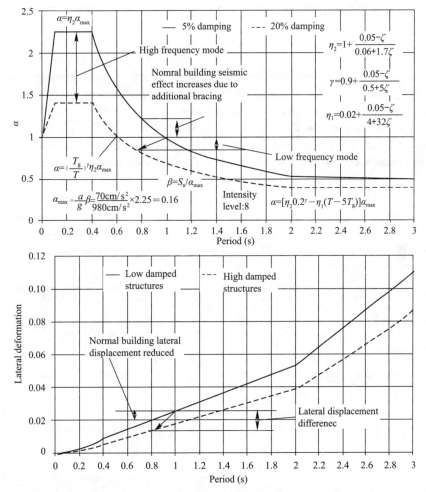

Fig. 9-5 Principle of energy dissipation

2. Advantages and Applications of Energy Dissipation Technology

Energy dissipating structural systems have several advantages compared to traditional seismic structural systems.

(1) Reliable. Compared with traditional seismic structures, the dynamic response of energy dissipating structures can be reduced by 40%—60%.

(2) Economical. In energy dissipating structures, it is not necessary to increase member sections and augment reinforcing bars to improve seismic behavior, leading to cost savings.

(3) Reasonable. In traditional seismic structures, the seismic action increases with the improvement in strength and stiffness, and this deficiency restricts the development of super high-rise buildings and large-span structures. For energy-dissipating structures, an increase in the structural height, flexibility or span is beneficial to the energy dissipation effect based on the principle that energy is dissipated by specific devices when large deformations occur.

Currently, energy dissipation technology is widely applied in several types of structures, such as high-rise buildings and super high-rise buildings, high-rise flexible structures and high-rise towers, long-span bridges, flexible pipelines and reinforcement engineering of existing buildings for earthquake or wind resistance.

9.3.2 Energy Dissipation Devices

An energy-dissipating structural system consists of the primary structure and energy-dissipating members or devices, which can be classified into several types based on the construction exterior and energy dissipation form.

1. Classification Based on the Construction Exterior

Based on the construction exterior, energy-dissipating members can be classified into several types:

(1) Energy-dissipating bracings. As a substitute to common structural bracings, energy-dissipating bracings can play a role in both lateral stiffness and energy dissipation. Square frame bracings, round frame bracings, cross bracings, diagonal bracings and K-shaped eccentric bracings are all energy-dissipating bracings, as indicated in Fig. 9-6.

(2) Energy-dissipating shear walls. As a substitute for common structural shear walls, energy-dissipating shear walls can play a role in both lateral stiffness and energy dissipation. Energy-dissipating slits between the walls and the energy-dissipating materials filled in walls help the shear walls to dissipate energy. Vertical slit shear walls, transverse slit shear walls, diagonal slit shear walls, surrounding slit shear walls, viscoelastic material-filled shear walls and viscous material-filled shear walls are all energy-dissipating shear walls, as indicated in Fig. 9-7.

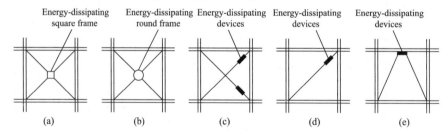

Fig. 9-6 Various energy-dissipating bracings
(a) Square frame bracings; (b) Round frame bracings; (c) Cross bracings; (d) Diagonal bracings; (e) K-shaped eccentric bracings

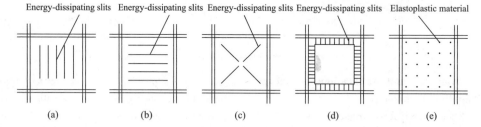

Fig. 9-7 Various energy-dissipating shear walls
(a) Vertical slit shear walls; (b) Transverse slit shear walls; (c) Diagonal slit shear walls;(d) Surrounding slit shear walls;
(e) Integrate shear walls

(3) Energy-dissipating joints. Energy-dissipating devices are installed at the beam-column joints (depicted in Fig. 9-8), and these devices start working when certain angle changes occur.

(4) Energy-dissipating connections. Energy-dissipating devices at the slits of the structures or connections of the structural members start working when relative deformations occur at the slits or connections.

(5) Energy-dissipating supporting or overhanging members. Various supporting or overhanging energy-dissipating devices can be installed on pipeline structures.

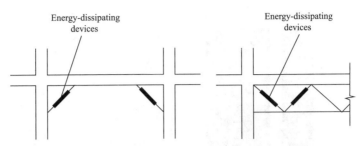

Fig. 9-8 Energy-dissipating beam-column joints

2. Classification Based on the Energy Dissipation Type

Devices are characterized by their force-deformation behavior, as provided in Table 9-1.

The introduction to a few energy-dissipating devices is given below.

(1) Metallic dampers

Metallic dampers dissipate energy by yielding metal pieces. An hourglass-shaped steel plate, which resembles the distribution of the stress in the cross section, allows for the distribution of plastic strains throughout the plates due to bending. Fig. 9-9 illustrates the details of metallic dampers.

Classification of energy dissipation devices   Table 9-1

Displacement-activated	Velocity-activated	Motion-activated
Metallic dampers	Viscous dampers	Tuned-mass dampers
Friction dampers		
Self-centering dampers		
Mixed type: Viscoelastic dampers		

Fig. 9-9 Metallic dampers

(2) Buckling restrained braces (BRB)

Unlike a normal brace, a BRB is a steel brace encased in an unbounded concrete tube to prevent buckling. Details of the BRB and its seismic performance are briefly depicted in Fig. 9-10.

(3) Viscous dampers

Viscous dampers provide a velocity-dependent damping force by forcing fluid through an orifice to dissipate energy.

The force-velocity relation can be linear or nonlinear and given by:

$$F_d = C_v \text{Sgn}(V) |V|^\alpha \qquad (9-3)$$

where,

F_d——the damping force;

C_v——the damping coefficient;

V——the velocity;

α——the exponential coefficient.

Details of the buckling restrained brace are briefly depicted in Fig. 9-11.

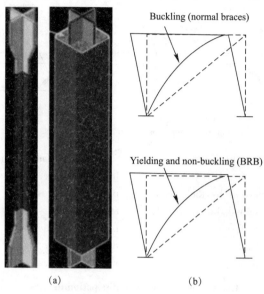

Fig. 9-10 Buckling restrained braces
(a) BRB details; (b) Seismic performance

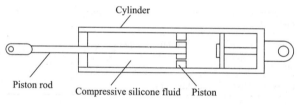

Fig. 9-11 Buckling restrained brace

(4) Viscoelastic Dampers

Viscoelastic dampers are combinations of viscoelastic materials and steel plates, as indicated in Fig. 9-12. This type of damper dissipates energy by the hysteretic shear deformation of viscoelastic materials.

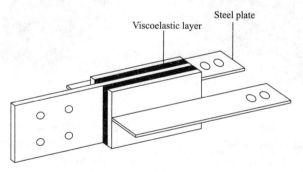

Fig. 9-12 Viscoelastic dampers

9.3.3 Applications

Recently, massive experimental and theoretical studies have been conducted on the reliability and durability of existing dampers, development of new dampers, restoring force models of dampers and methods of analysis and design for energy-dissipating structures. A few examples of the application of energy dissipation technology in buildings are provided here. The World Trade Center in New York, which was destroyed by a terrorist attack in September 2001, installed approximately 10,000 viscoelastic dampers in each tower, and these dampers performed extremely well even in bad weather. In 1992, viscous damping walls were applied in the SUT Building in Shizuoka, Japan, and judging from the analysis data, the viscous damping walls can reduce the seismic response to an extent of 70%—80%. The Jinghua Shopping & Leisure Center in Taipei is the first building that is installed with a triangular plate damping and stiffness device and has exhibited good seismic performance.

The following images are representative applications in real-world structures (Fig. 9-13).

Fig. 9-13 Applications of the energy dissipation technology

References

[1] Ministry of Housing and Urban-Rural Development of the People's Republic of China. Code for Seismic Design of Buildings (GB 50011—2010) Version 2016 [S]. Beijing: China Architecture and Building Press, 2016.

[2] Ministry of Housing and Urban-Rural Development of the People's Republic of China. Code for Design of Concrete Structures (GB 50010—2010) Version 2015 [S]. Beijing: China Architecture and Building Press, 2015.

[3] Ministry of Housing and Urban-Rural Development of the People's Republic of China. Code for Design of Masonry Structures (GB 50003—2011) [S]. Beijing: China Architecture and Building Press, 2011.

[4] Ministry of Housing and Urban-Rural Development of the People's Republic of China. Standard for Design of Steel Structures (GB 50017—2017) [S]. Beijing: China Architecture and Building Press, 2017.

[5] Ministry of Housing and Urban-Rural Development of the People's Republic of China. General Code for Seismic Design of Buildings and Municipal Engineering (GB 55002—2021) [S]. Beijing: China Architecture and Building Press, 2021.

[6] National Standardization Management Committee. The Chinese Seismic Intensity Scale (GB/T 17742—2020) [S]. Beijing: China Quality and Standards Press, 2021.

[7] National Standardization Management Committee. Seismic Ground Motion Parameters Zonation Map of China (GB 18306—2015) [S]. Beijing: China Standards Press, 2015.

[8] Hu Y.. Earthquake Engineering [M]. Beijing: Earthquake Press, 2006.

[9] Moehle J. P.. Seismic Design of Reinforced Concrete Buildings [M]. New York: McGraw-Hill Education, 2015.

[10] Paulay T., Priestley M. J. N.. Seismic Design of Reinforced Concrete and Masonry Buildings [M]. New York: John Wiley & Sons, Inc, 1992.

[11] Zhou Y.. Seismic Resistance of Building Structures [M]. Wuhan: Wuhan University of Technology Press, 2021.

[12] Lu X., Xiong H.. Seismic Resistance of Building Structures [M]. Beijing: Higher Education Press, 2019.